T0304671

# Corrosion Induced Failure of Bridge Post-Tensioned Tendons

This book provides an overview of post-tensioning (PT) in reinforced concrete bridge structures. It specifically addresses corrosion issues that were initially unforeseen but subsequently occurred, raising concerns regarding structural integrity and long-term viability.

It begins by exploring the introduction and background of PT bridge development, addressing post-tensioning systems and the various types of PT bridge construction methods. It also covers corrosion issues and failures that have previously occurred, and other designs for durability. Further, it goes into the details of experimental investigations of corrosion-caused bridge PT tendon failures, and the modeling efforts of wire and strand fractures. Lastly, it provides an assessment of the path forward, suggesting alternatives for current inspection and condition assessment.

This book is highly recommended for graduate and postgraduate students of civil, structural, and mechanical engineering. It will also be an interesting read for structural and civil engineers, and researchers involved with integrity assurance of PT structures.

# Corrosion Induced Failure of Bridge Post-Tensioned Tendons

William H. Hartt and
Teddy S. Theryo

CRC Press
Taylor & Francis Group
Boca Raton   London   New York

CRC Press is an imprint of the
Taylor & Francis Group, an **informa** business

A BALKEMA BOOK

Designed cover image: William H. Hartt and Teddy S. Theryo

First published 2025
by CRC Press/Balkema
4 Park Square, Milton Park, Abingdon, Oxon, OX14 4RN

and by CRC Press/Balkema
2385 NW Executive Center Drive, Suite 320, Boca Raton FL 33431

CRC Press/Balkema is an imprint of the Taylor & Francis Group, an informa business

*British Library Cataloguing-in-Publication Data*
A catalogue record for this book is available from the British Library

ISBN: 9781032357225 (hbk)
ISBN: 9781032357232 (pbk)
ISBN: 9781003328193 (ebk)

DOI: 10.1201/9781003328193

Typeset in Times New Roman
by Newgen Publishing UK

**Disclaimer:** The opinions, findings, recommendations, and conclusions expressed
in this book are those of the authors and not necessarily those of the Florida
Atlantic University (FAU), Florida Department of Transportation (FDOT), and BCC
Engineering, USA.

This book is dedicated to our family members who have tirelessly supported us during the writing of this book

To
My wife Sarah and my three children (who are no longer children): Tolly, Will, and Jennifer

To
To my wife Maria and my two children Grayce & Peter Theryo, including the late of Mgr. Nicolas Pierre Van der Westen, SSCC who supported me to become an engineer

# Contents

*Foreword*                                                                    ix

*About the authors*                                                           xi

1  Introduction and background                                                1

2  Post-tensioning hardware, bridge types, and construction
   methods                                                                    8

3  Corrosion-related bridge post-tensioned tendon failures                   18

4  Design for durability                                                      30

5  Experimental investigations of corrosion-caused bridge
   PT tendon failures                                                        37

6  Modeling and projection of corrosion-related bridge PT
   tendon failures                                                           45

7  Inspection, condition assessment, and remediation options                 77

8  The path forward                                                          91

*Index*                                                                      93

# Foreword

This book by William Hartt and Teddy Theryo is a definitive account of the durability issues confronting those responsible for structures containing post-tensioning tendons. Professor Hartt has been at the forefront of research and investigations into the problems of post-tensioning tendons in bridges in the United States since they were first discovered and investigated. Teddy Theryo's work in Florida has had to address the issues in a state with some of the most severe chloride exposure and, hence, corrosion risks leading to extensive problems with corrosion of prestressed bridge elements in the United States .

The inherent problem with ducted internal post-tensioned segmental bridges in particular is the difficulty of inspection through ducts and also of repair. This particularly applies to older bridges that have experienced the most problems to date. The book leads the reader through, first, the occurrence of the first failed tendons identified in Europe; second, the national and international collaboration that resulted in addressing these problems; and, lastly, the latest research, particularly Professor Hartt's own work to elucidate the mechanisms and to predict the extent of tendon failure.

Post-tensioned structures need to be buildable, inspectable, and maintainable. Extensive research has been carried out on ways of inspecting for corroded tendons and how to control tendon corrosion with multiple requests for proposals from the Federal Highway Agency, the Transportation Research Board, and others. Unfortunately, the nature of a grouted strand in a duct makes it extremely difficult to characterize or control corrosion, as shown here by the limited options discussed compared to the wide range of inspection and remediation techniques available for inspecting and repairing conventionally reinforced concrete structures.

While improvements have been made to the design and construction of post-tensioned structures, time will tell as to how effective they are in controlling tendon corrosion. Even if no problems are found with the latest cohort of designs, there remains a large inventory of existing bridges and other post-tensioned structures that will suffer from the problems described in this book;

and those responsible for these structures will need the guidance provided. This book should be read by all those involved in the design, inspection, and maintenance of structures containing post-tensioned tendons.

John P. Broomfield, D. Phil. (Oxon), EurIng, FICorr,
FIMMM, FNACE, FCS, CSci
Institute of Corrosion Level IV, Senior Cathodic Protection
Engineer, NACE International
Corrosion Specialist, Broomfield Consultants,
Consulting Corrosion Engineers

# About the authors

**William H. Hartt** served for 40 years as Professor of Ocean Engineering and Director of the Center for Marine Materials at Florida Atlantic University. During that time and subsequently, he directed numerous research projects and served as a consultant. He has authored approximately 100 journal publications pertaining to corrosion, fracture, and failure of engineering materials.

**Teddy S. Theryo** is a former Major Bridge Design Engineer with Florida Department of Transportation during writing of this book. Since February 2024 he is employed by BCC Engineering in Florida, United States. He has over 47 years of professional experience and has served in various capacities in construction, inspection, design, research, repair, and durability investigations of major post-tensioned concrete bridges around the world. He graduated from Gadjah-Mada University, Yogyakarta, Indonesia, in civil engineering and MSCE in structural engineering from the University of Washington, Seattle, Washington, United States and is a licensed professional engineer in the states of Florida and California. His specialty is in the areas of post-tensioned prestressed concrete design of major complex bridges, including segmental concrete and cable-supported bridges. He is an active member of several national (ASBI, PTI) and international technical committees (fib). Recently, he was awarded a Fellow status of fib.

# Chapter 1

# Introduction and background

Bridges, being inanimate objects, do not receive the same interest or attention as historical monarchs, kingdoms, or wars; however, for the technical mind with a bent toward the past, they pose immense interest. In general, bridges consist of, first, horizontal (or nearly so) surfaces that experience vertical loads from dead weight and traffic and, second, supports that sustain vertical compression. There are seven generalized categories for bridges according to design, construction, and historical progression, as listed in Table 1.1, the selection of which depended, or now depends, upon service requirements and the terrain or waterway being traversed. The simple beam approach dates to prehistoric times as simply a tree log spanning a narrow body of water or marshland. The arch bridge extends to ancient times, for example, the Arkadiko Bridge in Greece, which is thought to be the oldest example that remains in use today.

The arch bridge served as a breakthrough accomplishment for the ancients. Such structures make use of the fact that materials of construction at the time such as stone, while weak in tension, are strong in compression. Given its configuration, the arch structure inherently transfers loads from the superstructure through compression members to the substructure, although requiring that the end footings be secure. The arch then supports the deck superstructure. There are magnificent, ancient examples of this construction that remain today as illustrated, for example, by the Saint Angelo Bridge in Rome which dates to the early second century. Critical to such construction was the development and use of cofferdams whereby water could be walled off and a solid foundation placed. Such construction and also the road-building technology that the Romans at the time developed were key to their ability to sustain the empire for as long as they did.

A second category of early arch bridge-type structures is aqueducts. A surviving example is the Pont de Gard in southern France, which dates to the early first century. Through this bridge, water flowed along a 50-km long elevated pathway to the Roman colony at Nimes.

I

DOI: 10.1201/9781003328193-1

*Table 1.1* Listing of bridge types

- Simple beam
- Arch
- Truss
- Segmental concrete bridges
- Extradosed
- Finback / sail bridge
- Cable stay
- Suspension

*Figure 1.1* Photograph of the Zhaozhou Bridge (photostock.com.cn).

Also constructed were what have been termed clapper bridges, which translates from Latin as "pile of stones" or Anglo-Saxon as *cleaca*, which means steppingstones. These bridges employed relatively long, thin stone slabs that served as the deck with piers comprised of large rocks or stones. A classic example that remains today is the Postbridge in Devon, England, which dates to medieval times.

In addition, noteworthy historical bridges remain in China, as exemplified by the 1,400-year-old Zhaozhou Bridge. Here, the deck is supported by open arch spandrels that extend from an arch. Figure 1.1 provides a photograph. Menn has published a comprehensive, historical account of the progression of bridge types, designs, and materials of construction.[1]

The evolution and transition to present-day wider bridges with longer spans, reduced materials of construction, a more streamline appearance, and the capacity to accommodate greater loads and traffic, were facilitated by the development of improved cements and then concrete and inclusion of iron and ultimately steel as reinforcement. The latter (steel reinforcement) is particularly noteworthy since it has high tensile strength compared to concrete, which is weak in tension, thus facilitating designs that would otherwise not be possible. The categories include, first, steel frame or truss construction, second, conventionally reinforced concrete, which employs carbon steel reinforcing bars with surface deformations conforming to ASTM A615,[2] and, third, pre- and post-tensioned (PT) concrete, which utilizes seven-wire, spirally wound, cold-drawn, pearlitic steel strands conforming to American Society for Testing and Materials (ASTM) Standard A416 with a Guaranteed Ultimate Tensile Strength (GUTS) of 1860 MPa (270 ksi).[3] While strands are most common, individual wires and bars are also employed in some instances.

The concept of post-tensioning was first introduced and patented by the French engineer Eugene Freyssinet in 1939; and the concept has since evolved to its present technological state. Also, Freyssinet invented and obtained a patent in France for a conical friction post-tensioning anchorage for 12 wires as shown in Figure 1.2. Because of this, the year 1939 is considered the birth of

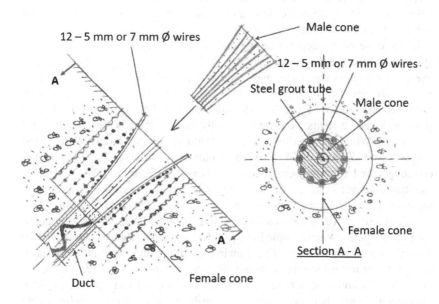

*Figure 1.2* Design drawing for the Freyssinet cone friction PT anchorage.

post-tensioning worldwide.[4] Although the relatively simple Freyssinet conical-shaped high-strength concrete anchorage is no longer used, it has inspired other PT system types. These have evolved over the years with different materials, shapes, and sizes for PT wires, strands, and bars to what are presently employed. Freyssinet was also responsible for innovations that pertain to prestressed concrete.[5,6]

Figure 1.3 shows a schematic illustration of bridge types that have been made feasible because of these more modern concepts and materials of construction. As such, the truss design utilizes an all-steel frame construction. Span by span and modern cantilever constructions are invariably dependent upon post-tensioning for structural integrity, while suspension and cable-stay bridges, in addition to use of high-strength wires and cables, are likely to utilize PT in some elements.

Bridge PT tendons can be either bonded or unbonded, where the former exhibits strain compatibility in strands and concrete at the ultimate limit state, but for unbonded it is not. Bonding is affected by grout that fills the duct cavity after strand stressing, whereas unbonded tendons typically employ micro-crystalline wax, grease, or gel within the duct, which precludes such bonding. European bridge designs have employed unbonded tendons for some years, as have nuclear power containment vessels in the United States. However, in response to integrity issues for bonded tendons, as discussed subsequently, recent bridge projects in the United States have also been completed that employ unbonded tendon technology, although building construction industry employed unbonded tendons for some 70 years.

Irrespective of bonded versus unbonded, tendons can be either internal or external, where the former are embedded in concrete and the latter are not, except at deviators and anchorages, as explained subsequently. Chapter 2 provides more detail regarding the various types of PT bridge designs and their construction methods. Typically individual wires and bars can be employed in post-tensioned elements. Seven-wire spirally wound strand is most common for PT bridges. While diameter for Grade 1860 strands (1,860 MPa (270 ksi) GUTS) ranges from 9.53 mm (0.375 in.) to 15.24 mm (0.600 in), the latter is invariably employed for PT bridge construction as shown in Figure 1.4. The applicable standard does not address wire chemistry but, instead, dimensions and strength alone; however, chemical composition is invariably near that of the iron–carbon eutectoid reaction (0.77 wt% carbon), which results in essentially a fully pearlitic microstructure (microscopic $Fe_3C$ platelets in an $\alpha$-Fe (body-centered cubic) matrix) and, along with the wire drawing operation and resultant work hardening, serves to provide the requisite strength. Typically, 4, 7, 12, 15, 19, 29, or 31 strands comprise a tendon, although this number can vary. Also, there have been studies that investigated the feasibility of employing high-strength austenitic and duplex stainless steels for prestressed marine pilings.[7–9] These

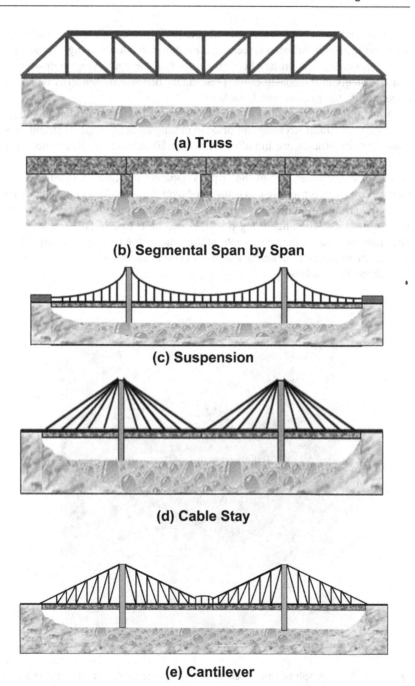

**(a) Truss**

**(b) Segmental Span by Span**

**(c) Suspension**

**(d) Cable Stay**

**(e) Cantilever**

*Figure 1.3* Schematic illustration of modern bridge designs.

have involved ranking of alloys according to performance in various accelerated corrosion tests and included screening for environmental cracking. Apparently, little or no consideration has been given to-date for employing high-strength stainless steels in PT construction, presumably because of concerns regarding susceptibility to environmental cracking.

Further, Figure 1.4 provides a view of an in-place bridge PT tendon anchorage subsequent to strand stressing but prior to cutting of the extending strand tails. Bridge tendon strands are initially stressed to 80 percent of GUTS; however, this reduces to about 70 percent upon insertion and seating of wedge grips and load release (the outer face of the grips is apparent in Figure 1.4). Strand tails are then cut, and a protective cap is installed over the outer face of the anchorage assembly. Grouting is then performed via tubes that typically extend from the duct to the deck. The underlying principle and objective of post-tensioning is placement of tendons in such a configuration that, once tensioned, compressive stresses are imparted to concrete that would otherwise be in tension and subject to cracking and spalling.

*Figure 1.4* Photograph of the exterior of an in-place bridge tendon subsequent to stressing, insertion of wedges, and load release. (www.concrete network.com/post-tension).

While post-tensioning can be classified as structural concrete, the technology differs markedly from construction that employs conventional reinforcement alone. In structural concrete, deformed bars of often greater diameter but lower strength than PT strands are utilized. Also, while PT strands are stressed in the long-term, after concrete creep and shrinkage and strand relaxation, to 60 to 63 percent of GUTS, conventional reinforcement is stressed only by dead weight and service loadings. Consequently, the same amount of corrosion on a wire as on a bar can cause fracture of the former but not the latter. Also, corrosion of the conventional reinforcement results in solid reaction products, the specific volume of which exceeds that of the unreacted iron. Because concrete is brittle and of low tensile strength, these expansive reaction products result in concrete cracking and spalling. While this, in and of itself, can result in costly repairs and structure rehabilitations, still the damage is visible and can be detected and addressed in a timely fashion. This is not the case for PT strands, however, which are typically embedded in grout, encased in a duct, and for internal tendons embedded in concrete. Consequently, integrity assurance for PT bridges presents different, more complex challenges than for conventionally reinforced ones. Subsequent chapters discuss this point in greater detail.

## REFERENCES

1  C. Menn, *Prestressed Concrete Bridges*, Birkhäuser Verlag AG, Basel, Switzerland, 1990.
2  ASTM A615/A615M-20, "Standard Specification for Deformed and Plain Carbon-Steel Bars for Concrete Reinforcement," American Society for Testing and Materials, West Conshohocken, PA, 2020.
3  ASTM A416/A416M-18a, "Standard Specification for Steel Strand, Uncoated Seven-Wire for Prestressed Concrete", ASTM International, 100 Barr Harbor Drive, West Conshohocken, PA, 2018.
4  P. Xercavins, D. Demarthe, K. Shushkewich, "Eugene Freyssinet – His Incredible Journey to Invent and Revolutionize Prestressed Concrete Construction", 3rd fib Congress, Washington D.C., 2010.
5  P. Xercavins, D. Demarthe, and K. Shushkewich, 3rd fib International Congress, 2010, p. 1.
6  M. Sanabra-Loewe and J. Capella-Liovera, *PCI Journal*, Fall, 2014, p. 93.
7  J. Sanchez, J. Fullea, and C. Andrade, *Advances in Construction Materials*, Part V, 2007, p. 397.
8  Y. Wu and U. Nurnberger, *Materials and Corrosion*, Vol. 60, 2009, p. 771.
9  J. Fernandez, A. A. Sagüés, and G. Mullins, "Investigation of Stress corrosion Cracking Susceptibility of High Strength Stainless Steels for Use as Strand Material in Prestressed Concrete Construction in a Marine Environment," paper no. 2686 presented at CORROSION 13, NACE International, Houston, 2013.

# Chapter 2

# Post-tensioning hardware, bridge types, and construction methods

## POST-TENSIONING SYSTEMS

The accomplishments of Freyssinet, as discussed in Chapter 1, have inspired the development of modern-day PT systems that evolve in different materials and designs for PT wires, strands, and bars, such as those developed by BBRV, Freyssinet, DSI, VSL, CCL TENSA, and others. These PT systems are proprietary hardware products that consist of a bearing plate, trumpet, wedge plate (anchor head), grout cap, and grout port or ports. In response to corrosion issues that have been disclosed and which are discussed in subsequent chapters, the designs of PT anchorages have been modified over time in order to better facilitate inspection and, if needed remediation.

Historically, anchorages employed but a single port located at the top of the anchorage for filler injection. However, as of 2003, the Florida Department of Transportation (FDOT) in the United States has required an additional vertical grout port or vent positioned at the top of the trumpet that facilitates borescope inspection subsequent to filler injection of any voids that are disclosed along with a permanent anchorage cap, as shown in Figure 2.1.

## BRIDGE POST-TENSIONING TENDON TYPES

Post-tensioned bridges have historically consisted of grouted internal and external tendons. However, bridge projects in European countries and the United States have recently been completed that employ flexible fillers, as mentioned in Chapter 1, and which are discussed in greater detail in Chapter 4. Internal tendons are cast into and bonded to the structural concrete, as illustrated schematically in Figure 2.2; and both types are housed in either corrugated metal (grouted tendons), high-density polyethylene (HDPE), or polypropylene (PP) ducts. However, while corrugated galvanized ducts have been widely employed in the past, they are generally no longer permitted for corrosive environments because of corrosion issues. The high-strength steel that comprises tendons can be strands, wires, or bars; however, strands are the most prevalent,

DOI: 10.1201/9781003328193-2

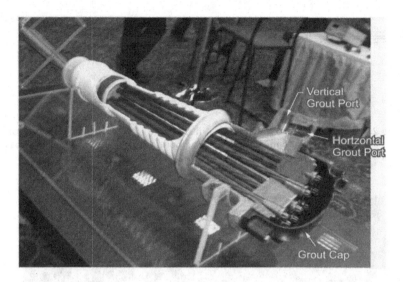

*Figure 2.1* Photograph of a PT anchorage system as modified (DSI System).

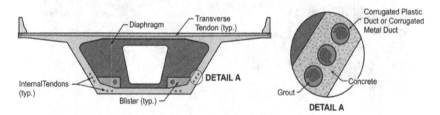

*Figure 2.2* Schematic cross-section illustration of internal PT tendons in a bridge superstructure segment.

as described and discussed in Chapter 1. External tendons, on the other hand, are not embedded in or bonded to the concrete structural section except at deviators and anchorages, as shown by the schematic illustration in Figure 2.3. Figures 2.4 and 2.5 are the photographs of external tendons at a deviator and diaphragm, respectively. A recent development of external tendons is that they are no longer bonded in a deviator or diaphragm but pass through a diabolo form. The benefits of a diabolo form are ease of installation, larger placement tolerance, and simpler tendon replacement should this become necessary.

Typical tendon types employed in post-tensioned bridges include the following:

1. Top longitudinal internal cantilever tendons
2. Continuity top and bottom flange internal tendons

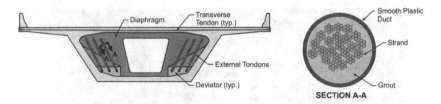

*Figure 2.3* Schematic cross-sectional illustration of external PT tendons in a bridge superstructure segment along with a tendon cross-section view.

*Figure 2.4* Photograph of grouted external and bonded tendons in a deviator.

3. Continuity longitudinal draped external tendons
4. Transverse internal tendons in the top flange
5. Vertical internal tendons in the webs
6. Vertical internal tendons in diaphragms
7. Transverse internal tendons in diaphragms.

## PT BRIDGE CONSTRUCTION METHODS

There are presently multiple post-tensioned bridge construction methods in the industry, including segmental bridge construction, along with advancements

*Figure 2.5* Photograph of grouted external tendons and bonded in a diaphragm.

in the design of mechanized/automation erection equipment and heavy lifting capabilities. The objectives of these modern post-tensioned bridge construction methods are to improve efficiency and speed of construction, while maintaining quality and safety at an optimum level. Some of the most common construction methods are described below. The construction method selected dictates the design strategy and the type of tendons employed. For instance, with cast-in-place PT structures, the primary tendons required are draped ones with a profile that balances positive moments. For a balanced cantilever construction, the cantilever (top) tendons are the primary tendons needed to balance the negative moments during erection. The application of post-tensioning technology to a variety of bridge construction methods is described below. A more detailed coverage of PT bridge construction methods is presented in literature references.[1–3] A summary of these is presented below:

1. *Cast-in-Place (CIP) PT Box-Girder Bridge on False Works*: A bridge girder is a horizontal, longitudinal structural member that supports the superstructure and thereby resists vertical loads, whereas false works are temporary constructions analogous to scaffolding that support an elevated girder during construction. Bridges of this type consist of single or multi-cell box girders, T beams, or slabs. Typically, such bridges have internally draped continuous tendons in the webs from end to end of the girder

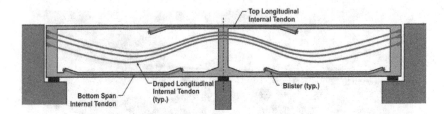

*Figure 2.6* Schematic elevation view of the tendon layout in box girders.

(Figure 2.6). Some tendons may be anchored at intermediate diaphragms. For bridges of a long span or spans, additional internal tendons are also provided in the top and bottom flanges and anchored in what are termed blisters, as shown in Figures 2.6 and 2.7. The top deck can be either transversely post-tensioned or conventionally reinforced. However, a transversely post-tensioned deck is preferred.

2. *Cast-in-Place Segmental Balanced Cantilever Bridge*: Cast-in-place segmental box-girder bridges are commonly employed for long spans constructed using the balanced cantilever method with a set of form travelers. For these, successive segments that are approximately 4.5–5 m (15–16 feet) long are cast against the previously post-tensioned segment. These bridges employ internal cantilever tendons in the top flange over the webs during construction. After the free cantilever is connected with

*Figure 2.7* Photograph of a blister in top flange of a box girder.

closure CIP concrete, continuity tendons are stressed in the bottom and top flanges, the latter being anchored at blisters, as shown in Figure 2.7. These may be supplemented by additional external draped tendons from diaphragm to diaphragm. The top deck is typically post-tensioned transversely employing tendons encased in flat ducts. In the case of very long span bridges, it is also common to employ internal vertical PT bars in the webs of segments that are close to a pier for shear design enhancement.

3. *Precast Segmental Balanced Cantilever Bridge*: Erection of bridges in this category employs the balanced cantilever method using either an overhead gantry, a beam and winch, a segment lifter, or a ground-based crane. Figure 2.8 provides an example of a segment lifter for segment erection. The individual segments are match-cast in a short or long line casting yard in about 3–4 m (approximately 10–12 feet) lengths. During erection, epoxy is applied on both faces of the segment match-cast joints with integrity being affected by internal cantilever tendons in the top flange.

4. *Precast Segmental Span-by-Span Bridge*: This category of bridges consists of precast match-cast segments, approximately 3–4 m (about 10–12 feet) long, erected using an under-slung or overhead gantry. All spans are temporarily supported on a gantry and are stressed together using temporary PT bars after epoxy is applied on the match-cast joints. The CIP

*Figure 2.8* Precast balanced cantilever construction with a segment lifter.

joints are cast between precast segments and the pier segments. Permanent longitudinal external tendons are post-tensioned from both diaphragms to complete the span construction, and the process is repeated at the next adjacent span.

5. *Full-Span Bridge Construction*: The full-span construction method is suitable for short-span PT box girder bridges. The box girder is cast in the casting yard for the whole span and then transported over the completed span using a specialized equipment that serves as a transporter and erector as shown in Figure 2.9. This construction method is economical for very long total bridge lengths with uniform short spans. The recent construction of high-speed rail elevated superstructures has successfully employed this method.

6. *CIP on Movable Scaffold System* (MSS): The MSS construction method involves casting-in-place the whole superstructure span on a movable scaffold steel truss system, as shown in Figure 2.10. The subsequent segment/span is cast against the previous segment/span and the two are post-tensioned together. Then the MSS system is moved forward ready for the next span. This construction method is very similar to CIP on falseworks.

*Figure 2.9* Full-span erection method (photo courtesy of Deal).

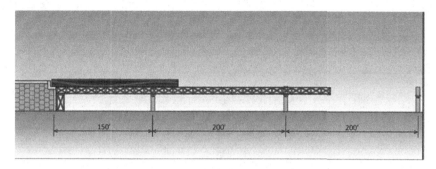

*Figure 2.10* Schematic illustration of the movable scaffold system erection method.

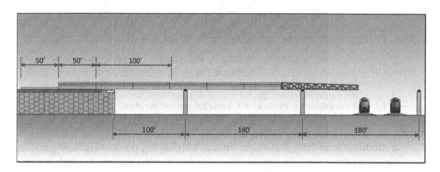

*Figure 2.11* Schematic illustration of the incrementally launched method.

7. *Incremental Launching Bridge*: The concept of the incremental launching method is shown in Figure 2.11. This consists of a stationary short casting bed that is located on one of the bridge abutments. The first segment is placed, the subsequent segment is cast against the first, and post-tensioning is applied. A temporary steel launching nose is attached to the first segment. The segment and launching nose are pushed or pulled against the abutment forward using hydraulic jacks until the launching nose reaches the other side of abutment. Typically, the superstructure is designed as a continuous structure. During erection, uniform post-tensioning is needed, and after the superstructure is in place, draped tendons either bonded or unbonded are installed to handle permanent and live loads. The advantage of this erection method is that the operation is performed over the pier columns without temporary support on the ground.

8. *Precast Spliced I Girder Bridges*: Bridges of this type have been constructed increasingly within the past several decades in situations where

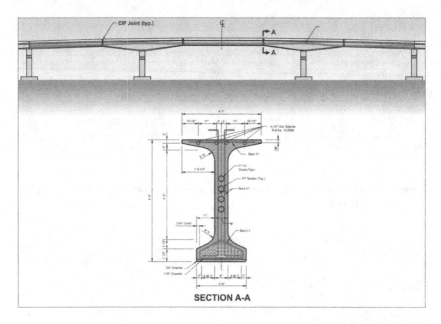

*Figure 2.12* Schematic of a typical PT spliced girder bridge.

a medium span length is required and employing precast I-shape girders instead of box girders. Other cross-section shapes such as U girders also have gained popularity. Figure 2.12 shows a schematic illustration of a PT spliced girder bridge. Here, several long pretensioned I-girders or bulb-tee girders are post-tensioned together using draped internal tendons in the web, thus forming a continuous multi-span girder end to end. The joints between girders consist of cast-in-place concrete closure pours. The diaphragms are typically cast at the splice locations and are either conventionally reinforced or transversely post-tensioned. During erection, temporary supports are provided at the CIP joints to provide stability until the girders are made continuous. The tendon types employed in this construction are, first, draped longitudinal internal tendons and, second, transverse internal tendons in the diaphragms at cast-in-place joints.

## REFERENCES

1. Teddy S. Theryo, William H. Hartt, and Piotr Paczkowski, "Guidelines for Sampling, Assessing, and Restoring Defective Grout in Prestressed Concrete Bridge Post-Tensioning Ducts" FHWA Publication No.FHWA-HRT-13-028, Washington, D.C., October 2013.
2. Walter Podolny, Jr. , and Jean M. Muller,., *Construction and Design of Prestressed Concrete Segmental Bridges*, Wiley, 1982.
3. Jacques Mathivat, , *The Cantilever Construction of Prestressed Concrete Bridges*, Wiley, 1983.

Chapter 3

# Corrosion-related bridge post-tensioned tendon failures

## HISTORY OF CORROSION-RELATED BRIDGE PT TENDON FAILURES

The initial occurrence of a bridge PT tendon corrosion issue came to light in 1967 with the collapse of the Bickton Meadows Footbridge in Hampshire, England. This was followed in 1985 by the collapse of a small, 32-year-old bridge in West Glamorgan, UK, because of tendon corrosion at a construction joint.[1,2] Subsequently, in 1992, the River Schelde Bridge in Belgium collapsed due, at least in part, to corrosion of tendons. Also, corrosion-caused bridge tendon failures has been reported from France. The occurrence of these failures led to inspection of PT bridges in numerous countries, and these revealed instances of tendon corrosion due primarily to inadequate protection being afforded to strands by the grout. Consequently, in 1995 the Transportation Research Agency in the UK, Service d'Etudes Technique des Routes et Autoroutes (SETRA), and Laboratoire Central des Ponts des Chausees (LCPC) in France conducted a joint study of durability issues associated with bridge PT tendons, as discussed in detail in Chapter 4.

In 1999, the Florida Department of Transportation (FDOT) discovered efflorescence at anchorages, water leakage at joints, and grout void issues on six PT bridges. This was followed three years later by disclosure during a routine biannual inspection of a failed external tendon on the Niles Channel Bridge in the Florida Keys.[3] The failure occurred slightly inboard of an anchorage and was attributed to the presence of either bleed water or water entry into a grout void from an external source. Three years subsequent to the Niles Channel Bridge tendon failure, two failed tendons were discovered on the seven-year-old, 5.8 km (3.6 mile) Mid Bay Bridge that crosses the Choctawhatchee Bay near Destin, Florida. These failures resembled that of the Niles Channel Bridge. Of the 846 tendons, an additional 11 tendons were replaced because of corrosion. Subsequently, in 2019, eight additional tendons with corrosion on this bridge were replaced, these being identified as at risk during a routine biannual inspection. The finding that two of these tendons were adjacent to one another on

DOI: 10.1201/9781003328193-3

the same span resulted in bridge closure and emergency repairs. A combination of causes, including segregated/carbonated grout, cracked ducts, and excessive bleed water, were identified as responsible.

Additional PT bridges in the United States with identified tendon corrosion issues as of the year 2021 include the following:

1. Roosevelt (Stuart), Selmon West Extension (Tampa), and Channel 5 (Florida Keys) Bridges in Florida.
2. Clive Avenue Bridge in Indiana, which was closed in 2009 after 26 years in service because of issues with corroded tendons and rebar that were determined to be beyond repair.
3. Varina Enon, Lord Delaware, and Eltham Bridges in Virginia for which tendon corrosion was attributed to improper grouting.
4. Plymouth Avenue Bridge in Minnesota, where corrosion of internal tendons as a consequence of water and deicing salt leakage that resulted from misaligned drainage piping was disclosed at a bridge age of 27 years.
5. Wando River Bridge in South Carolina on which a corrosion-caused tendon failure was found at age 30 years. This resulted in an extended closure.

The situation for the Roosevelt Bridge was particularly problematic in that, at age 23 years, five failed tendons were discovered in close proximity to one another on one side of the bridge resulting in its shutdown and immediate repairs. Tendon configuration in this case consisted of four to nine embedded tendons on each side of the span base. The bridge was constructed using precast segmental balanced cantilever however, several segments adjacent to abutment were supported on falsework. The corroded tendons were located at the closure pour segment at the end span. Other tendons also exhibited corrosion. The failures were attributed to water leakage at a closure pour and resultant corrosion of the galvanized ducts. The Wando River Bridge (item 5 above) also employed galvanized ducts. However, as noted earlier, many Departments of Transportation no longer allow galvanized ducts. Aliofkhazrael et al.[4] have recently reviewed the role of hydrogen in embrittlement of grouted PT tendons with galvanized ducts.

e FDOT has estimated that, as of 2011, approximately $55·million had been spent on the repair of 11 PT bridges in that state.[5] This figure could only have gone up in the interim as additional instances of corrosion and failure have been disclosed. Tendon failures have also been reported on bridges in Europe, as noted above.[6] Such issues can only be expected to increase in number as these structures continue to age.

In response to the above failures and concerns regarding tendon corrosion, as well as their own experiences, multiple European countries, including the

United Kingdom, France, Germany, and Italy, as well as Japan and South Korea have performed comprehensive inspections of PT bridges and reported corrosion issues and, in some cases, tendon failures. Deficiencies or factors of concern that were identified include (1) inadequate design and construction practices, (2) age, (3) poor quality, segregated or deficient grout, (4) grout voids, and (5) presence of free water, sulfates, and chlorides.

Research studies have addressed an improved understanding of, first, grouts for post-tensioning,[7] second, the relationship between the extent of corrosion and residual wire/strand strength,[8] and, lastly, susceptibility of strands to hydrogen embrittlement.[9] In an attempt to address failures that have been caused by grouting issues, as was the case for many of the above bridges, the industry subsequently transitioned to prepackaged, thixotropic grouts with components being proportioned at a production facility. As such, these should not be subject to errors that might arise from job site proportioning other than possibly for mixing water. Such grouts, being thixotropic, become less viscous under pressure and thereby tend to better fill tortuous void spaces. However, while this change in grout type may have addressed one problem, it has given rise to another with the occurrence of what has been termed "soft grout"; that is, grout that is locally pliable, moist, of relatively low chlorides and high pH, but with elevated sulfates. The latter species (sulfates) has been shown to be a potential cause for steel corrosion.[10,11] Soft grout has been projected to result from a suspension of segregated, less dense particles in bleed water that rise to the top of the duct incline.[12] Consequently, a likely site for presence of soft grout and resultant strand corrosion is at high elevations along the profile of longitudinal draped tendons such as at anchorages and intermediate piers. For transverse tendons, the high point typically occurs at mid-position. Also, for draped tendons, strands are invariably pressed upon the top of ducts along horizontal runs between deviators and, as such, may not be fully embedded in grout.[13] It is also feasible that particulates comprising soft grout could become trapped on the underside of congested strands, particularly along horizontal runs between deviators.

The photograph in Figure 3.1 shows a cross-section view of a tendon from the Ringling Causeway Bridge in Florida at a location on a failed tendon that was fully grouted. On the other hand, Figure 3.2 shows a similar view of this same tendon at a location just inboard of an anchorage and adjacent to the site where it failed from corrosion. Various grout types that were encountered in the latter case are identified.[14] The material labeled "wet plastic grout" is soft grout. When this tendon was opened, free water flowed from the void area. Locations with segregated, chalky grout are also identified; however, the grout was sound elsewhere.

A second, albeit unrelated tendon grout issue arose in 2011 when a major manufacturer of prepackaged grout in the United States informed the Federal Highway Administration that the product produced over an eight-year period at

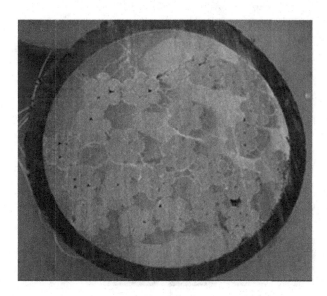

*Figure 3.1* Cross-section view of a tendon from the Ringling Causeway Bridge showing sound grout (Florida Department of Transportation).

one of its plants in the United States contained chloride concentrations ranging from below the specified limit (0.08 weight percent per cementitious material) to as high as 400 percent above this limit. It was determined that the cement manufacturer was recycling waste from the kiln, and with every recycle, the chloride concentration increased. No chemical analyses were performed for quality assurance that would have detected chloride. Some 200 bridges were identified as potentially affected. Bridge owners were notified regarding this finding, and a report was issued that provided guidance.[15]

Subsequently, chemical analyses of the grout in question from a bridge under construction in Texas revealed chloride concentrations in excess of 5 wt.%.[16] Figure 3.3 shows a photograph of an opened anchorage from this bridge, which reveals the presence of moist soft grout (A in Figure 3.2) and voids. Free water was also present, and corrosion of strands in the soft grout was apparent. As a consequence, tendons on this bridge that were fabricated using a specific grout lot that was identified as being at issue were replaced. Development of improved inspection techniques has been identified as the most critical issue for PT bridge integrity assurance.[17]

A report released by the U.S. Federal Highway Administration in 2002 indicated the annual cost of corrosion to the United States as of 1998 was $8.79 trillion or 3.1 percent of gross national product.[18] Of this, $22.6 billion was attributed to infrastructure and $8.3 billion of that to highway bridges. While

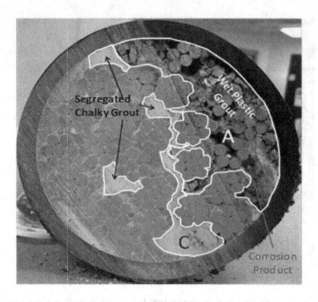

*Figure 3.2* Cross-section view of a failed tendon from the Ringling Causeway Bridge (Florida Department of Transportation).[22]

there are multiple deterioration mechanisms for concrete bridges, in the case of structures that are conventionally reinforced with bars or prestressed strands, chloride-induced corrosion that leads to resultant concrete cracking and spalling caused by expansive reaction products resulting from reinforcement corrosion is the most prevalent deterioration mechanism. For such structures, there are generally visual indications from established, standardized inspection and evaluation techniques that can define the extent of any such pending issue or issues. Irrespective of any concrete cracking and spalling, electrochemical techniques employing surface-mounted or embedded probes can distinguish between passive and active corrosion reinforcement,[19] in addition to measuring polarization resistance[20] as an indicator of corrosion rate. However, this is unlikely to be the case for PT bridges, since structural integrity is provided foremost by tendons rather than conventional reinforcement; and strands for these are encased in grouted ducts. As such, the components that are intended to provide corrosion protection also render inspection and condition assessment difficult at best. A recent Federal Highway Administration (FHWA) study reported that as of 2016, only three of some 353 segmental box PT girder bridges in the United States were structurally deficient.[21] By way of contrast, 1,523 of 5,872 (28 percent) of conventionally reinforced girder/floor beam bridges were so rated. A probable cause for this distinction is that corrosion-related deterioration

*Figure 3.3* Photograph of an opened tendon anchorage showing soft grout (A) and a void.[24]

issues are more difficult to detect for PT compared to conventionally reinforced bridges. Where corrosion-related issues have occurred on PT bridges, these have been attributed to aggressive environments, contractor inexperience or lack of understanding of the concepts involved, inadequate design details and supervision, and failure to correct identified problems. As such, comparison of the structural integrity/deficiency for PT versus conventionally reinforced bridges may be misleading. Thus, all bridges in the United States are typically subjected to a biannual inspection, which may be limited to a visual walkthrough. For PT bridges the only visually apparent indication of a deficiency is likely to be a sagging or separated tendon, as exemplified, for example, by the failed tendon on the Varina Enon Bridge. Figure 3.4 shows a photograph of this tendon. As such, there is no disclosure of a pending problem but only of failure after the fact. Also, such determinations can be made only for external and not internal tendons.

For a typical span-by-span PT bridge, of the type shown in Figure 2.5 with three draped tendons on each span side, losing two tendons on a given span reduces structural capacity by approximately one-third; however, if these two tendons are on the same side of the span, excessive deflection and concrete cracking would occur and the bridge would be closed for immediate repair. The probability of two tendon failures occurring on the same span may seem

*Figure 3.4* Photograph of a failed tendon on the Varina Enon Bridge.[21]

very improbable. However, consider the Ringling Causeway Bridge mentioned above in conjunction with Figures 3.1 and 3.2, which has 27 spans. For such a bridge with three tendons on each side, as shown in Figure 2.5 this then translates to 162 tendons total. The probability of a second failure occurring on the same span of such a bridge as an initial failure, assuming randomness, then is 0.03, and for such a failure to take place on the same side of the span as an initial failure 0.006. Both are very low probabilities. However, for this bridge (Ringling Causeway), two failed tendons and 15 others with corrosion were replaced some eight years post construction. Because all these tendons were in the same general vicinity on the bridge, apparently due to material or construction issues (or both), it is concluded that randomness cannot necessarily be assumed.

As noted above, corrosion protection for tendons is critical for the structural integrity of PT bridges. The concrete structure itself, in which tendons are positioned, serves as an initial means for so providing, perhaps more so for internal compared to external tendons. This is followed by the duct in which strands are positioned. Third, there is the cementitious grout, which, if compositionally sufficient, physically sound, and of high pH, promotes passivity and corrosion resistance for strands embedded therein. Also required is protection for grouting tubes, which typically terminate at the deck and can be subject to physical damage during construction and, consequently, to water intrusion. There have been several reported instances of this having occurred.[22,23] In one case, the contractor was required to grout tendons within six months of strand tensioning, a deadline that was not always met. However, when grouting was to commence and a protective cap was removed from an anchorage on one tendon, water was observed pouring from within the duct. The Post-Tensioning Institute published a specification for grouting PT structures in 2001, and there have been several subsequent additions.[24] Lastly, for satisfactory performance anchorages must provide protection at strand terminations, including within the critical region of the wedge grips. While grout provides protection to the interior anchorage face, assuming there are no voids at this relatively high elevation on the profile of draped tendons, protection beyond this within the wedge grips is physical. Corrosion within grips has been identified as an issue that can cause the release of wires or an entire strand and loss of prestress along a development length.[25]

## THE CORROSION PROCESS

Corrosion is an electrochemical process, meaning that it involves both charge transfer and chemical change. Prestressing steel consists predominantly of a heavily cold-drawn pearlitic microstructure, as discussed in Chapter 1, that is, of microscopic cementite ($Fe_3C$) platelets in a ferrite (body-centered cubic) iron matrix. Differences between the two phases and even differences within the same phase give rise to variations in electron affinity; that is, to strength of the bond of valence electrons to the atom core. Consequently, there is a thermo-dynamic driving force to reduce these differences; and this results in electro-chemical half-cell reactions at different micro- or macro-sites on the wire/strand surface upon exposure to an electrolyte. The latter can consist of any aqueous phase capable of ionic charge transport, such as pore water within grout or free water within grout voids. These reactions are:

$$Fe \rightarrow Fe^{+2} + 2e^- \quad \text{(anodic sites) and} \tag{3.1}$$

$$H_2O + \tfrac{1}{2}O_2 + 2e^- \rightarrow 2OH^- \quad \text{(cathodic sites).} \tag{3.2}$$

For reinforcing steel in conventional concrete, loss of steel section is normally not as critical as for PT wires and strands, since there is no prestress and the resultant expansive stresses caused by the relatively high specific volume corrosion products ($Fe(OH)_2$, $Fe_3O_4$, and $Fe_2O_3$) result in visible concrete cracking and spalling which, while significant maintenance wise, provide ample time for condition assessment and repair/rehabilitation. Irrespective, steel is normally passive and exhibits a low corrosion rate in alkaline media such as cementitious pore water. This passivity is compromised, however, by presence of chlorides or sulfates in sufficient concentration, resulting in localized attack (pitting). Crevices, such as are formed where strand wires contact one another, give rise to differential aeration cells that constitute another form of localized attack, where steel within the deaerated crevice is anodic to grouted regions for which the pore water is of high pH and is aerated. Figure 3.5 shows a photograph of localized pitting on a wire and Figure 3.6 of separated, corroded strand wires, both recovered from the grout void region of a mock-up tendon after six months

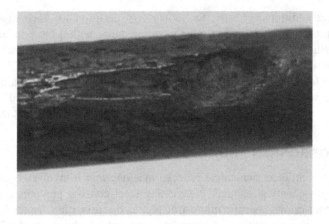

*Figure 3.5* Photograph of a pitted strand wire recovered from near the grout void interface of a mock-up tendon with zero admixed chlorides subsequent to cleaning (pit depth 2 mils (0.05 mm)).[26]

*Figure 3.6* Photograph showing the corrosion morphology of wires from the grout void region of a mock-up tendon.[26]

exposure.[26] Relatedly, Figure 3.7 shows a photograph of a wire that exhibits crevice corrosion along what was the line of contact with another wire.

In addition, a unique contributor to corrosion was disclosed on the Mid-Bay Bridge mentioned above, where duct cracking occurred, as shown in Figure 3.8.[27] This cracking was typically at the 10 and 2 o'clock orientations and was apparently caused by bending stresses in the duct about a grout–void interface that resulted from thermal contraction when temperature dropped. This, in turn, facilitated moisture/water access to strands at these locations.

*Figure 3.7* Photograph of a strand wire with crevice corrosion along the line of what was contact with another wire.[26]

*Figure 3.8* Photograph of a cracked duct from the Mid-Bay Bridge.[27]

Together, the above examples of wire and strand fractures and tendon failures, where these have occurred, define a formidable challenge for the transportation community. While these have been addressed to some extent by enhancements in materials, designs, and monitoring practices, as well as greater awareness by transportation officials of potential PT bridge corrosion issues, still a major challenge moving forward will be integrity assurance for the vast inventory of earlier vintage PT bridges, not only in the United States, but throughout the developed world.

## REFERENCES

1  R.J. Woodward and F.W. Williams, "Collapse of Ynys-y-Gwas Bridge, West Glamorgan, *Proceedings of the Institute of Civil Engineers*, Vol. 84, 1988, p. 635.
2  R.J. Woodward, "Collapse of a Segmental Post-Tensioned Concrete Bridge," *Transportation Research Record*, Vol. 1211, 1989, p. 38.

3   R.G. Powers, A.A. Sagues, and Y.P. Virmani, "Corrosion of Post-Tensioned Tendons in Florida Bridges," Proc. 17th U.S.-Japan Bridge Engineering Workshop, Japan Public Works Research Inst. Technical Memorandum No. 3843, p. 579, 2002,

4   M. Aliofkhazrael, D. Dukeman, and C.L. Alexander, *Materials Performance*, Vol. 61(3), 2022, p. 48.

5   S-K Lee, *Materials Performance*, Vol. 51(2), 2012, p. 40.

6   L. Bertolini and M. Carsana, "High pH Corrosion of Prestressing Steel in Segregated Grout," *Modeling of Corroding Concrete Structures*, Eds. C. Andrade and G. Mancini, RILEM Series 5, Cedex, France, 2011.

7   A.J. Schokker, J.E. Breem, and M.E. Kreger, *Materials Journal*, Vol. 98, 2001, p. 296.

8   R.G. Pillai, P. Gardoni, D. Trejo, M.B.D. Hueste, and K.F. Reinschmidt, *ASCE Journal of Materials*, Vol. 22(10), October 2010.

9   J. Fernandez and A.A. Sagüés, "Embrittlement Susceptibility of Stressed High Strength Carbon Steel Ungrouted Post-Tensioning Strands", paper no 7757, Corrosion 2016, NACE International, Houston, 2016.

10  V. Gouda, *British Corrosion Journal*, Vol. 5, 1970, p. 198.

11  L. Bertolini and M. Carsana, "High pH Corrosion of Prestressing Steel in Segregated Grout," RILEM Book Series 5, Rilem, Cedex, France, 2011.

12  H.R. Hamilton, A. Piper, A. Randell, and B. Brunner, "Simulation of Prepackaged Grout Bleed under Field Conditions," Final Report submitted to Florida Department of Transportation, Gainesville, FL, April 2014.

13  M. Sprinkel, *PTI Journal*, July 2015, p. 1.

14  K. Lau, I. Lasa, and M. Parades, *Materials Performance*, November 2013, p. 64.

15  T.S. Theryo, W.H. Hartt, and P. Paczkowski, "Guidelines for Sampling, Assessing, and Restoring Defective Grout in Prestressed Concrete Bridge Post-Tensioning Ducts," FHWA Report No. FHWA-HRT-13-028. Federal Highway Administration, Washington, D.C., October 2013.

16  B.D. Merrill, "Grout Testing and Analysis," Texas DOT Memorandum, September 14, 2010.

17  "Transportation System Preservation Research, Development, and Implementation Roadmap," Federal Highway Administration, Washington, DC, January 2008.

18  "Corrosion Costs and Preventive Strategies in the United States", Federal Highway Administration Report No. FHWA-RD-01-156, Federal Highway Administration, Washington, DC, 2002.

19  "Standard Test Method for Corrosion Potential of Uncoated Reinforcing Steel in Concrete," ASTM Standard C876-15, American Society for Testing and Materials, 100 Barr Harbor Drive, West Conshohocken, PA 15428.

20  "The Determination of Corrosion Rate of Steel Embedded in Concrete by the Polarization Resistance and AC Resistance Methods," STP 1183025, American Society for Testing and Materials, 100 Barr Harbor Drive, West Conshohocken, PA 15428.

21  Federal Highway Administration, Deficient Bridges by Superstructure 2016, www.fhwa.dot.gov/bridge/nbi/no10/strtyp16.cfm.

22  R.A. Reis, "Corrosion Evaluation and Tensile Results of Selected Post-Tensioning Strands at the SFOBB Skyway Seismic Replacement Project Phase III Report," California Department of Transportation, Sacramento CA, September 5, 2007.

23  C. Roepke and R. Huet, "Post-Tensioned Tendon Failure Analysis at the Hawaii Rail Transit Project," Exponent, Inc. Doc. No. 1604126.000-1306, July 2016.

24  PTI M55.1, "Specification for Grouting of Post-Tensioned Structures," Post-Tensioning Institute, Farmington Hills, MI, 2001.

25 W.H. Hartt, J.F. Vincent, and V.I. Ivanov, "Wire and Strand Slippage – A Potential Failure Mode for Post-Tensioning Tendons," *Materials Performance*, Vol. 54(10), 2015, p. 34.
26 S-K Lee and J. Zielske, "An FHWA Special Study: Post-Tensioning Tendon Grout Chloride Thresholds," Report No. FHWA-HRT-14-039, Federal Highway Administration, Washington, DC, April 2014.
27 W.H. Hartt and S. Venugopalan, "Corrosion Evaluation of Post-Tensioned Tendons on the Mid-Bay Bridge in Destin, Florida," Final Report to Florida Department of Transportation, Tallahassee, FL, 2002.

# Chapter 4

# Design for durability

In response to collapse in the Ynys-y-Gwas Bridge in the UK and the River Schelde Bridge in Belgium (see Chapter 3), as well as disclosure of PT tendon failures in France, a high priority was placed upon reviewing the design, detailing, and construction practices in place at that time to determine the root cause or causes of these occurrences. The identified PT bridge durability issues were common to both France and the UK. France is the pioneer in the application of prestressed concrete construction, and the UK has been credited with the subsequent development and innovation of prestressed concrete structures. In addition, both countries accumulated significant experience in design, construction, repair, and management of post-tensioned concrete. The UK Highway Agency, Transport Research Laboratory, Service d'Etudes Technique des Routes et Autoroutes (SETRA), and Laboratoire Central des Ponts des Chausees (LCPC) in France jointly initiated in January 1995 a study of post-tensioned bridge durability issues. The main objective of this collaboration was to explore the evolution of post-tensioned bridges in each country and seek solutions to durability issues. The results of these studies were published in 1999 as a book titled, *Post-Tensioned Concrete Bridges*.[1] Prior to this UK–France collaboration, private industries in the UK published through the Concrete Society the first edition of Technical Report No.47, "Durable Post-Tensioned Concrete Bridges."[2] This contained recommendations for improved design and detailing of ducts, grouting systems, and grout materials, in addition to certification of post-tensioning operations along with training and testing. The second edition of TR 47 was published in 2002 and also addressed external unbonded tendon construction, remedial grouting of existing bridges, and new test methods. The Concrete Society subsequently published TR 72 in 2010 as a replacement of TR 47 Second Edition.[3] This was an improved revision of TR 47 Second Edition and was expanded to include the international community, such as Federation international du beton (fib), the United States, and other European countries. It also addressed grouting and unbonded tendons in buildings. The UK works are significant in that they recommend not only multiple layers of corrosion

DOI: 10.1201/9781003328193-4

protection but also improved construction practices, design, and detailing of new structures. The TR 47 recommendations were credited with the decision to lift the ban for cast-in-place PT bridges in the UK, although this remained in place for precast segmental bridges with internal tendons. Apparently, concern for the authority was that the duct connection at match-cast joints of precast segmental construction had not been satisfactorily addressed with regard to corrosion protection. Thus, without a special duct coupler, the ducts are disconnected at the match-cast joint where only a thin layer of epoxy is applied. This type of detail does not meet the multi-level protection criterion (explained below), although a segmental duct coupler at match-cast joints has been widely used subsequently in the United States.

The study of PT durability issues initiated by the UK–France collaboration was continued by the fib and International Association for Bridge and Structural Engineering (IABSE) as the 1st Durability of Post-Tensioning Tendons Workshop, held on November 15–16, 2001, in Ghent, Belgium.[4] This workshop included international delegations that reported PT durability issues encountered in their own countries, including Japan, Germany, the United States, Switzerland, and Canada, as well as the UK and France. Keynote speakers were selected from countries with PT durability issues and they described the status of PT bridges, inspection reports, approaches for repair, and strategies for improvement. PT durability issues have led to restrictive use by some local authorities, such as the UK Department of Transport (now Highways Agency). In the United States, the Florida Department of Transportation (FDOT) adopted a new approach termed "New Direction for Florida's PT Bridges."[5] This consisted of five elements, as listed below:

1.  Enhanced post-tensioned systems
2.  Fully grouted tendons
3.  Multi-level anchor protection
4.  Watertight bridges
5.  Multi-tendon paths.

Element No.2 was modified in January 2016 upon approved use of flexible fillers in PT tendon ducts as an alternative to cementitious grout. This was the first workshop of its kind where international experts were together, learned from each other, and formulated a strategy for moving forward. One of the objectives was to find a long-term solution to overcome PT durability issues. Research projects had been and were being conducted at that time, especially in European countries and Japan, in order to understand the root cause(s) of the corrosion issues. It was determined that there were multiple common durability issues internationally, but there were also durability issues specific to each country. Some of these are listed as follows:

1. Grout voids at high elevations along ducts
2. Grout segregation
3. Blockage during grouting
4. Presence of water in ducts
5. Corroded steel tendons ranging from minor to complete failure
6. Ungrouted PT ducts
7. Insufficient concrete cover over tendons
8. Lack of PT anchorage sealing
9. Leaking expansion joints.

Post-tensioning corrosion protection options and procedures such as choice of grout material, grout injection method, grout testing, and construction quality control and quality assurance were extensively discussed. The durability defects that resulted in the Ynys-y-Gwas Bridge collapse, which occurred without warning, were only small in number and not encountered by all countries. The presence of grout voids in PT ducts at high elevations and other locations along PT tendons was common among the countries. The challenges shared by all countries were, first, difficulty in detecting grout voids in internal tendons and, second, replacement of defective tendons. In the workshop, it was apparent that each country had its own strategy for coping with PT durability issues. Of particular interest was how Germany was dealing with grout issues. First, quality assurance of the grout injection process was improved by establishing national and international quality control requirements for specialty contractors who install both internal and external tendons. Second, poor construction practice was countered by requiring controllable, exchangeable, and strengthened PT tendons. These strategies resulted in a preference for external rather than internal tendons. The German requirements for new PT bridges are as follows:

1. Webs must not include internal PT tendons.
2. Promotion of a factory-made corrosion protection system employing wax/grease and PE sheathing of individual strands (mono-strand system).
3. PT tendons must be controllable, restressable, and exchangeable, and
4. Increase the amount of conventional reinforcement to control cracking before rupture.

The fib/IABSE Second Workshop on Durability of Post-Tensioned Tendons took place in Zurich, Switzerland, in October 2004[6] as a continuation of the First Workshop that was held in Ghent, Belgium. The objectives in this workshop included a broad discussion of the fib draft report, "Durability Specifics for Prestressed Concrete Structures – Durability of Post-Tensioned Tendons" (March 26, 2004) and approval of resolutions on a recommended practice for post-tensioned tendons. This document later became fib Bulletin 33, "Durability of Post-Tensioning Tendons," which was published in 2006.[7,8] The workshop was divided into three sessions as:

Session 1: Design Concepts for Durable Post-Tensioning Tendons
Session 2: Materials and Construction
Session 3: Maintenance, Assessment, and Rehabilitation.

This Second Workshop stressed the point that durability begins at the conceptual design stage. As such, it is not sufficient just to improve the protection of PT tendons alone but instead to provide an overall design approach. The aggressiveness of the environment where the structure is to be located must be recognized and robust protection for tendons in extremely aggressive environments specified. In particular, polymer ducts must be used for extremely aggressive to even less aggressive environments. Listed below are critical factors that must be considered for post-tensioned structures:

1. Recognize that post-tensioned structures are not maintenance free as previously assumed.
2. Provide inspection and maintenance access at the design stage for critical elements such as PT anchorage areas.
3. Provide multiple levels of PT corrosion protection.
4. Provide ample of concrete cover.
5. Provide a dry environment around PT tendons, for instance by using waterproof sealing of the deck.
6. Provide provisions whereby additional tendons can be added in the future.
7. Consider redundant tendon systems. For example, it is better to have a greater number of smaller tendons than fewer larger ones.
8. Consider ease of tendon replaceability; that is, employ a design strategy that combines internal and external tendons.
9. Consider using a grout filler in combination with a flexible filler. Thus, use grout filler for straight tendons in the top deck and flexible filler for draped and external tendons.

The lessons learned from the above workshops have been listed in fib Bulletin 33. The fib strategy consists of three elements:

1. Design concept for durable post-tensioning tendons
2. Materials and construction
3. Maintenance, assessment, and rehabilitation.

This fib global strategy is very similar to that defined by the fib Second Workshop on Durability of Post-Tensioned Tendons, and the fib Model Code for Concrete Structures 2010[9] has implemented design specifications that stress the importance of conceptual design for durability. It is expected that national worldwide specifications would mimic and implement recommendations from the fib Model Code.

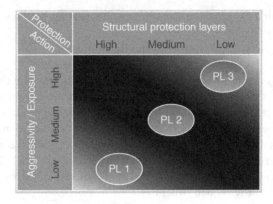

*Figure 4.1*  Fib protection level strategy. (Reproduced from fib Bulletin 33 with permission from fib.[7]).

A tendon protection strategy is one of the most important parts of the fib document, "Durability of Post-Tensioning Tendons," the cover of which is shown in Figure 4.1.

Here, aggressivity of the environment (AE) is ranked from low to high as are structural protection layers (PL). Also, detailed implementation guidance should be specified by local national codes. The fib recognizes three protection layers, which are designated as PL1, PL2, and PL3. A short description for each PL is given below:

PL1:  Bare PT strands or bars with corrugated metal duct and grout. A permanent gout cap is optional.

PL2:  Bare PT strands or bars with polymer / PE duct with grout/flexible filler plus permanent anchorage caps (fully encapsulated).

PL3:  PL2 with monitoring and ability to inspect the protection function at any time.

PL1 describes the minimum protection level required for PT bridges located in non-/low-aggressivity environments with high structural protection layers, such as inside a box girder.

PL2 describes the level of tendon protection required for PT bridges located in low to high aggressivity areas with a combination of from low to high structural protection, respectively as shown in Figure 4.1; for example, low AE plus low SP, medium AE plus medium SP, and high AE plus high SP fall into PL2.

PL3 represents the most stringent level of protection for PT bridges located in an extremely aggressive environment with very low structural protection. The required monitoring and inspection currently can only be met with the Electrically Isolated Tendon (EIT) method.[10] Using this method, the strands and

*Figure 4.2* PT anchorage EIT system (courtesy of VSL International).

anchorages of a tendon are electrically isolated from other metallic components, including reinforcing steel, such that any breach of the PE duct and potential for strand corrosion can be identified by electrical resistance measurements.

EIT method was developed in Switzerland in the 1990s as additional protection to PL2. PL2 is in principle a protection of tendon by full encapsulation for the whole length of tendon. The verification of leak tightness required is done by pressure test. PL3 is also full encapsulation, but the verification is by electrical resistance measurement. The steel tendon is isolated from the anchor head and bearing plate anchorage body by installing EIT plate as shown in Figure 4.2. The electrical monitoring can be done by an EIT wire connected to the tendon/anchor head and reinforcing bars around the tendon. Any defect of the duct, isolation plate, permanent caps, grout tubes / ports can be detected at any time during the life service of the tendon by measuring the electrical impedance resistance in the wire. One of the benefits of PL3 for instance is the ability to detect duct damage during concreting or grout tube/port connections defect. Stray current for railway bridge is another example where PL3 is an ideal tendon protection. PL3 can be applied for grouted internal or external tendon, as well as flexible filler for internal and external tendons.

Specifically emphasized by the fib strategy is that the level of corrosion protection required for bridge PT tendons must be determined during the conceptual design stage with the details evolving to final design. Another essential strategy from the owners' perspective is to treat PT hardware as a system from anchorage to anchorage instead of separate components consisting of bearing plate, anchor head, permanent anchorage cap, duct, duct couplers, inlets, outlets, vents, and duct fillers. Therefore, a pre-approved system and continuous quality assurance monitoring are needed. As a system, the integrity and quality of a PT system can be maintained from project to project.

## REFERENCES

1  "Post-Tensioned Concrete Bridges", Highway Agency, SETRA, TRL, LCPC, Thomas Telford, London, 1999.
2  "Durable Post-Tensioned Concrete Bridges," Technical Report No.47, The Concrete Society, Camberley, UK, 2002.
3  "Durable Post-Tensioned Concrete Structures," The Concrete Society, Technical Report No.72, Camberley, September 2010.
4  "Durability of Pot-tensioning Tendons", fib Bulletin 15, IABSE, AIPC, IVBH, and fib. November 2001.
5  "New Direction for Florida's PT Bridges", Corven Engineering, Tallahassee, October 8, 2004.
6  "Durability of Post-tensioning Tendons" Second Workshop, fib, IABSE, COST, ETH, October 2004.
7  "Durability of Post-tensioning Tendons", fib Bulletin 33, Lausanne, 2006.
8  "Monitorable Post-Tensioned Tendons", FHWA Research Report DTFH61-14-D00048/49/50, Federal Highway Administration, Washington, DC.
9  "fib Model  Code for Concrete Structures 2010 (2013)" Ernst & Sohn, Berlin, Germany.
10  "Guidelines Measures to Ensure Durability of Post-tensioning Tendons in Structures", ASTRA/SBB AG, Berne, Switzerland, 2007.

# Chapter 5

# Experimental investigations of corrosion-caused bridge PT tendon failures

## GENERAL

Corrosion- related bridge tendon failures can be considered in one of two general categories, as defined below:

Case 1: Tendon failure within several years as a consequence of soft grout, exemplified by what transpired on the Ringling Causeway Bridge (Figure 3.4).

Case 2: Tendon failure after several decades due to either bleed water or water from an external source in voids of otherwise sound grout, exemplified by what transpired on the Niles Channel Bridge.

Case 2 situations are well understood in terms of a macrocell established between portions of strands exposed to both water in a grout void and to adjacent, high pH, aerated pore water in otherwise sound grout. Case 1 occurrences, on the other hand, have only been identified within the past decade or so; and the mechanism of occurrence and variables of influence remain under study. Efforts regarding Case 1 include the following:

1. Identification of soft grout as being pliable, moist, and of relatively high sulfate concentration. The latter species (sulfate) has been shown to enhance steel corrosion in pore water type environments, as noted in Chapter 3.[1]
2. Investigation of tendon grout repair options. For situations where corrosion is relatively minor, this has involved the replacement of deficient grout with preapproved materials.[1]
3. Exposure of relatively small grout specimens prepared using material intended to promote occurrence of soft grout. Test variables have included prolonged time of grout storage, elevated storage temperature, and excess mix water.
4. Exposures employing a modified inclined tube test (MITT).[2,3]

 DOI: 10.1201/9781003328193-5

The MITT procedure has the advantage over small specimens (item 3 above) of replicating the geometry, elevation gradient, and macrocell activity that can be present in actual tendons. Also, any effect of strand congestion can be simulated and investigated by employing multiple strands within the duct. Macrocell activity, on the other hand, can be studied by supplementing strands with individual, instrumented steel specimens positioned along the tube incline. Also, the grout pumping stage of specimen preparation simulates what transpires in placement of actual tendons. Figure 5.1 shows a photograph of two MITT specimens under test.

## FLORIDA DEPARTMENT OF TRANSPORTATION (FDOT) SPONSORED RESEARCH

Randell et al.[4] conducted a MITT-based study for the purpose of identifying the cause(s) of soft grout. Experiments included both plain and prepackaged grouts, the latter consisting of lots provided by each of six manufacturers. Soft grout was encountered in the case of plain grout but neither bleed water nor soft grout resulted for prepackaged grouts provided these were within the manufacturer's recommended shelf life and were mixed according to directions.

However, factors such as excess mix water, elevated temperature and humidity during storage, and prolonged storage time did promote occurrence of soft grout. Inert fillers were identified as promoting soft grout, and caution has been suggested regarding using products that include these fillers.[4] A follow-on

*Figure 5.1* Photograph of two MITT specimens under test.[5]

study[6] focused more explicitly upon the above factors; specifically, prolonged storage time and elevated temperature and humidity during storage and identified and quantified the detrimental modification of grout constituents that resulted. Also developed was a field screening protocol that was projected to qualify grouts onsite.

Permeh et al.[7,8] performed MITT exposures that involved inclined 5.57m (18.3 feet) long PVC ducts containing two unstressed seven-wire strands and five individual, instrumented steel probes positioned at different elevations along the tube length. The grout employed was expired beyond the manufacture's recommended shelf life, and mix water was in excess of recommendation by a minimum of 15 percent. Admixed sulfates ranged from 0 to 5.5 wt% cement. The most severe grout segregation and enhanced sulfate concentrations occurred at higher elevations along the tube incline, and greater corrosion current densities resulted for probes in deficient grout at these locations.

The European response to concerns regarding bridge PT tendon integrity has resulted in several standards and recommended practices,[9-11] in addition to the report discussed in this chapter[6,12-15] as has also been the case with professional organizations in the United States[12,13] Standardized tests for grout qualification address setting time, viscosity, wet density, fluidity, wick-induced bleed, pressure bleed, and pump and fill ability.

## VIRGINIA DEPARTMENT OF TRANSPORTATION (VDOT) RESEARCH

In response to disclosure of the tendon failure on the Verina Enon Bridge (Figure 3.6), VDOT initiated studies that developed criteria for qualifying PT grouts for new bridges in that state. These consisted of both laboratory and field tests, the latter including full-scale tendon mock-ups. Figure 5.2 shows a schematic illustration of a tendon mock-up that was employed as part of grout qualification for a PT bridge on U.S. 460 near the Kentucky state line.[14] In this case, two mock-up tendons were constructed with a different candidate high-performance grout being employed for each tendon. Figure 5.3 shows the inlet port of each of the two mock-ups subsequent to grout setting and removal of the end caps. The tendon on the left was fully grouted with sound material, resulting in this grout being selected for the project, whereas the one on the right exhibited a grout void as well as soft grout. Based upon this experience, VDOT instituted a requirement that contractors for new bridge projects construct and conduct a mock-up duct test for grout qualification. This can be waived, however, if the grout at issue and the contractor are the same as for a previously approved project.[14]

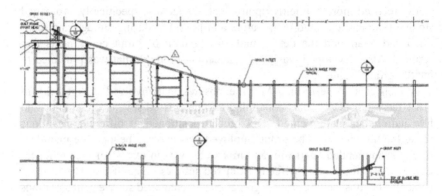

*Figure 5.2* Schematic longitudinal view of a tendon mock-up.[14]

*Figure 5.3* Photograph of the inlet end of each of two mock-up tendons subsequent to grout hardening and end cap removal.[14]

## FEDERAL HIGHWAY ADMINISTRATION (FHWA) RESEARCH

The FHWA also performed both laboratory bench-top-type experiments and simulated tendon exposures to better understand soft grout corrosion issues. Figure 5.4 is reproduced from Figure 8 of the resulting report[15] and provides a schematic, elevation view illustration of the load frame and specimen test

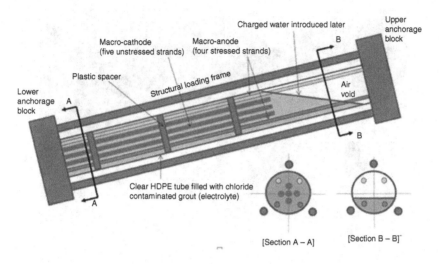

*Figure 5.4* Schematic illustration of the FHWA test frame. (Reproduced from Lee and Zielske[15].)

arrangement. This shows that the individual ducts contained four stressed strands with an intentional grout void space at the top of the incline in addition to four fully embedded unstressed strands, the latter serving as a macro-cathode; and Figure 5.4 provides a photograph. Grout-admixed chloride concentrations ranged from 0 to 2.0 wt%, and in total eight specimens were constructed and tested. Stressing was to 60 percent of GUTS, exposures extended for six months, and specimens were subsequently dissected, and strands evaluated. Terminations involved the application of an additional load to strands such that the wedge grips could be removed from the anchorage, and strands were then detensioned and recovered for analysis. Upon doing this, three wires on one strand of a tendon with 0.4 w/o Cl⁻ fractured near the grout–void interface. Figure 5.6 shows a photograph of the grout void region of this tendon with the fractured wires identified (red arrow). Also apparent is a black deposit on the grout void interface (gray arrows) which, upon chemical analysis, was determined to have a water-soluble sulfate concentration of 2.8 wt% cement. The wire fractures were attributed to a high corrosion rate that resulted from the sulfates. Figure 5.7 shows an enlarged, close-up view of this deposit and Figure 5.8 provides a close-up view of corroded strands at the grout void interface and Figure 5.9 of the fractured wires. Together, the above studies have contributed to a better understanding of factors and variables that affect bridge tendon corrosion and failure and of materials, design, and construction methodologies that better facilitate integrity assurance.

*Figure 5.5* Photograph of the full-scale tendon exposure specimens. (Reproduced from Lee and Zielske[15].)

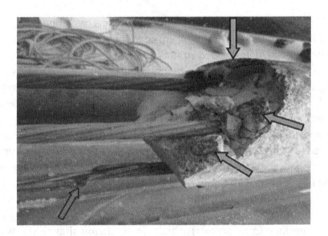

*Figure 5.6* Photograph of the grout void region of Specimen MS-0.4. The red arrow identifies the fractured wires. (Reproduced from Lee and Zielske[15].)

*Figure 5.7* Close-up photograph of corroded strands near the grout interface. Arrows indicate the black deposit. (Reproduced from Lee and Zielske[15].)

*Figure 5.8* Photograph of a strand with fractured wires. (Reproduced from Lee and Zielske[15].)

# REFERENCES

1  K. Lau, I. Lasa, M. Paredes, and J. Rafols, "Laboratory Corrosion Assessment of Post-Tension Tendons Repaired with Dissimilar Grout," paper No. 2602 present at CORROSION/13, NACE International, Houston, 2103.

2  J.P. Fuzier, *Durability of Post-Tensioning Tendons*, International Federation for Structural Concrete, Ghent, Belgium, 2001, p. 173.

3  R. Claussin and A Chabert, *Durability of Post-Tensioning Tendons*, International Federation for Structural Concrete, Ghent, Belgium, 2001, p. 235.

4  M. Randell, M. Aguirre, and H.R. Hamilton, *PTI Journal*, August 2015, p. 17.

5  H.R. Hamilton and G.A. Alverez, "Post-Tensioning Grout Bleed, Duct, and Anchorage Protection Test," Structures Research Report No. 929, Florida Department of Transportation, November 2002.

6  E. Torres, M. Aguirre, C.C. Fararo, and H.R. Hamilton, "Evaluation of Shelf Life in Post-Tensioning Grouts," Final Report submitted to Florida Department of Transportation, Gainesville, FL, August, 2018.

7  S. Permed, K.K. Krishna Vigneshwaren, M. Echeverria, K. Lau, and I. Lasa, *Corrosion*, Vol. 74, 2018, p. 457.

8  S. Permed, K.K. Krishna Vigneshwaren, K. Lau, and I. Lasa, *Corrosion*, Vol. 75, 2019, p. 848.

9  European Standard BS EN 445, "Grouting for Prestressing Tendons. Test Materials," British Standards Institute, 2007.

10  European Standard BS EN 446, "Grouting for Prestressing Tendons: Grouting Procedures," British Standards Institute, 2007.

11  European Standard EN 447, "Grout for Prestressing Tendons: Basic Requirements," British Standards Institute, 2007.

12  "Specification for Grouting of Post-Tensioned Structures," Post-Tensioning Institute, Farmington Hills, MI, 2007.

13  "Interim Statement on Grouting Practices," American Segmental Bridge Institute, December 2000. Buda, Texas.

14  M. Sprinkel, *PTI Journal*, July 2015, p. 1.

15  S-K. Lee and J. Zielske, "An FHWA Special Study: Post-Tensioning Tendon Grout Chloride Thresholds," Report No. FHWA-HRT-14-039, Federal Highway Administration, Washington, DC, April, 2014.

# Chapter 6

# Modeling and projection of corrosion-related bridge PT tendon failures

## BACKGROUND

In view of the bridge tendon corrosion and failure instances that have been reported till now, as described and discussed in previous chapters, and the criticality of tendons to PT bridge integrity, a methodology has been developed that relates the timing of such failures to wire corrosion rate statistics. The present chapter discusses how this modeling has evolved and the incremental enhancements that have been made. Model development, assumptions, and input parameters for the approach reflect forensic results from corroded PT wires and strands recovered from the FHWA-simulated tendons described in Chapter 5 that were tested under conditions considered representative of exposures that can occur in actual bridge structures, as described in conjunction with Figures 5.4 and 5.5. Upon termination of these tests, tendon disassembly and wire cleaning, corrosion morphology was visually assessed and the depth of pits determined. As an example, Figure 5.7 shows a close-up view of a wire with a shallow pit and Figure 5.8 the appearance of wires from the grout–void interface region of a tendon with soft grout but no admixed chlorides. It was assumed for projection modeling purposes that pits could be treated as advancing according to a planar front, as justified below.[1,2]

## TENDON FAILURE PROJECTION

The model that was developed projects the onset of wire and strand fractures and tendon failures and their progression with time.[2,3] Corrosion of tendon wires invariably involves, first, an initiation period and, second, a time of propagation; however, in the present approach, the former was assumed to be negligible. Justification for this considers that active corrosion is likely from the outset if soft grout or bleed water is present or is likely to be relatively short if water from an external source has access. Fundamental inputs to the modeling approach are, first, the distribution with exposure time of remaining cross-section area for individual wires, as represented by the mean and standard deviation of corrosion

DOI: 10.1201/9781003328193-6

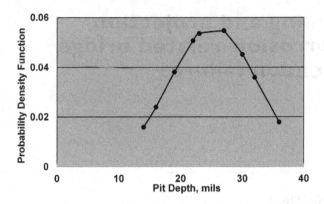

*Figure 6.1* Plot of the pit depth probability density function for the FHWA mock-up tendon with zero admixed chlorides (see Chapter 5).[1,3]

rate, expressed as μ(CR) and σ(CR), respectively; and second, the mean and standard deviation of residual strength for individual wires. An assumption of normality for the distribution of corrosion rates for individual wires was justified based upon the finding that maximum pit depth on individual corroded wires conforms to a bell-shaped trend,[1,3] as indicated by the data in Figure 6.1, for example. Considering then that stressing is to 63 percent of GUTS, which is representative of the long-term value in tendons subsequent to concrete creep and shrinkage and strand relaxation,[4] as discussed subsequently, it can be shown that once a third wire of a seven-wire strand fractures, then stress in the remaining four wires exceeds GUTS; and fracture of that strand is assumed. The number of strand fractures required to cause tendon failure depends, of course, on the number of strands comprising the tendon.

Information regarding the corrosion loss dependence of residual wire strength has been determined by an investigation that employed center (straight) wires that were precorroded subsequent to their removal from strands and then stressed to failure.[5,6] The results from this study are shown in Figure 6.2. The stress corresponding to 63 percent of GUTS (1,172 MPa), which is the upper limit of the range anticipated in the long-term subsequent to concrete creep and shrinkage and strand relaxation, is also indicated. The implications of a time dependence of stress decay in strands to the indicated long-term value are addressed subsequently. The equation for the best-fit line for the Figure 6.2 data was determined as

$$FS = 19.69 \cdot RA + 37.67 \ (R^2 = 0.93) \tag{6.1}$$

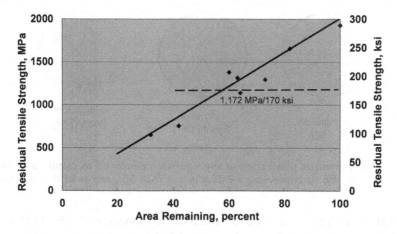

*Figure 6.2* Plot of residual strength of straight precorroded PT wires as a function of cross-section area at the most corroded location.

where FS is residual fracture stress (SI units) for remaining cross-section area and RA expressed as a percentage. While load from broken wires can be transferred to adjacent, unbroken ones, such occurrence was not incorporated into the modeling approach. Instead, wire fractures were assumed to transpire once cross-section was locally reduced to an extent that stress becomes elevated to the GUTS, irrespective of the condition of adjacent wires. Strand fracture was assumed to occur upon breakage of a third wire on that strand; and in this regard load transfer is reflected. That the strand in Figures 5.6 through 5.8 did not fracture subsequent to breakage of a third wire is attributed to the relatively short tendon length and the consequent likelihood that stressing was in displacement control. Also, this tendon was stressed to 60 percent of GUTS rather than the model assumed 63 percent. Equation 6.1 was modified for the modeling analyses such that FS at RA = 100 percent equals the GUTS (1,860 MPa (270 ksi)). Thus

$$FS = 19.69 \cdot RA - 107.42 \tag{6.2}$$

As noted above, PT wire corrosion can result from bleed water, charging of water from an external source into grout voids, or soft grout. Wire corrosion rate and resultant wire and strand fractures and, ultimately, the progression of tendon failures are determined accordingly. The modeling approach considers the distribution of corrosion caused cross-section loss for wires as a function of time with fracture of individual wires being projected from Equation 6.2. Because wires are assigned to strands, fracture of the latter results upon loss of a third wire, whereupon the GUTS is met or exceeded for remaining unfractured wires

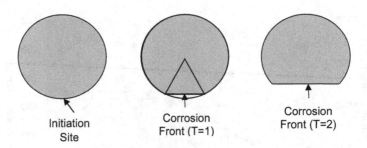

*Figure 6.3* Schematic illustration of the algorithm employed to calculate RCSA in the situation where RCSA is less than 50 percent of the original.

of that strand. The initial model consisted of 162 tendons, each with 22 Grade 270 seven-wire strands (3,564 strands and 21,384 outer wires total). Outer wires only are considered, given that the center wire is of a slightly larger diameter and receives some shielding from the outer wires. Such a tendon configuration was selected because it corresponds to that for the Ringling Causeway Bridge in Florida, two of which failed in less than eight years after construction as noted earlier; however, results of the analysis should be independent of the number of strands per tendon except for a possible rounding error. Irrespective of this, the model was subsequently modified and applied to tendons with 12, 15, and 19 strands per tendon as well.

This tendon failure projection modeling approach considers that corrosion is initiated at a single point on the wire circumference and subsequently propagates with time as a planar front at a rate determined by values for $\mu(CR)$ and $\sigma(CR)$. Accordingly, for a remaining wire cross-section area (RCSA) greater than 50 percent of the original, RCSA is calculated at successive time increments employing the algorithm shown schematically in Figure 6.3, whereby RCSA is determined at different time increments, according to the equation,

$$RCSA = A - SA + TA \tag{6.3}$$

where A is the original area, TA is the triangle area for which one side defines the corrosion front (shown in Figure 6.4 for T (time) = 1) and a later time (T = 2), and SA is the corresponding sector area. For RCSA values of 50 percent and less, the algorithm shown by Figure 6.4 and represented by the equation,

$$RCSA = 0.5 \cdot A - TRA(1) - TRA(2) - \ldots, \tag{6.4}$$

is employed, where TRA(-) is the area of successive approximated trapezoids the height of which depends upon corrosion rate with width the average of the

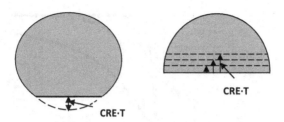

*Figure 6.4* Schematic illustration of the algorithm employed to calculate RCSA in a situation where RCSA is less than (a) and greater than (b) 50 percent of the original.

two corresponding cords. The total number of fractured wires is then determined at incremental times according to the following sequential steps:

1. Values for RCSA are determined at incremental times according to the above algorithm employing a specific choice for $\mu(CR)$ and values for corrosion rate ranging from $+2.20 \cdot \mu(CR)$ to $-2.00 \cdot \mu(CR)$ in 0.20 increments.
2. Fracture probability for each RCSA increment is calculated as the normal cumulative distribution function (cdf) of fracture stress using the mean and standard deviation of the fracture stress-remaining cross-section area data in Figure 6.2 (58.95 and 6.33 percent, respectively).
3. The number of fractured wires in each area increment is calculated for the same incremental times employed above for RCSA determinations as the product of the number of wires in each RCSA increment and the corresponding fracture probability.
4. Lastly, the total number of fractured wires is determined as the sum from all the increments.

Considering that the model, in this specific case, is based upon 162 tendons, each with 22 strands, as noted above, successive six random numbers of the 21,384 total are grouped and taken to represent the outer wires of individual, successive strands. The fracture of a given strand occurs then according to the time increment at which three of its wires are designated as fractured per the above protocol, since the assumed fracture stress (1,860 MPa (270 ksi)) is then reached or exceeded. Likewise, by the same rationale, the fracture of seven strands on a specific tendon results in the failure of the latter. Figure 6.5 shows example results based upon the original modeling protocol as a plot of the percentage of wire and strand fractures and tendon failures as a function of time for the specific case of $\mu(CR) = 0.8$ mmpy (3.0 mpy) and $\sigma(CR)$ 50 percent of $\mu(CR)$.

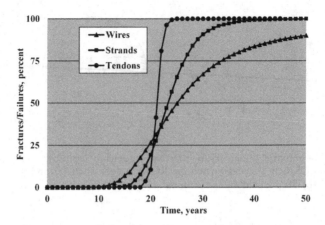

*Figure 6.5* Plot of the percentage of wire and strand fractures and tendon failures as a function of time for μ(CR) = 0.08 mmpy (3.0 mpy) and σ(CR) 50 percent of μ(CR).

However, the examination of the trend for wire and strand fractures relative to that for tendon failures in Figure 6.5 reveals an incompatibility, in that at year 25, as an example, all tendons are projected to have failed; however, not all wires and strands have fractured. This, of course, is an impossibility and occurred because the model, as initially developed, counted only wire fractures that resulted because of excessive section loss; however, once three wires of a strand fracture, the remaining four also fracture from tensile overload. Likewise, once seven of the 22 strands of a tendon fracture, so also do the remaining 15 for the same reason. Consequently, the modeling approach was refined to accommodate wire and strand fractures from both causes (excessive corrosion loss and tensile overload). Figure 6.6 shows the results of an analysis that incorporated wire and strand fracture from both causes employing the same corrosion rate statistics as above (μ(CR) = 0.08 mmpy (3.0 mpy) and σ(CR) 50 percent of μ(CR)). These results indicate that the trend for wire fractures is the same as in Figure 6.5 up to the time when strand fractures commence. Likewise, the trend for strands is the same until the onset of tendon failures. However, the trend for tendon failures is the same in both cases.

## MODEL VERIFICATION

Two opportunities arose whereby appropriateness of the model projections could be tested. The first resulted from fracture of wires on one of the FHWA mock-up tendons, where upon detensioning three wires fractured, as discussed in conjunction with Figures 5.6 and 5.8. These wire fractures were triggered

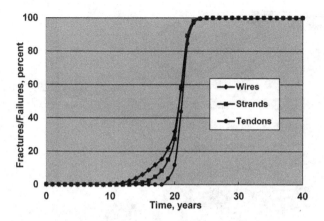

*Figure 6.6* Plot of the percentage of wire and strand fractures and tendon failures as a function of time for μ(CR) = 0.08 mmpy (3.0 mpy) and σ(CR) 50 percent of μ(CR) considering wire and strand fractures resulting from both corrosion loss and tensile overload.

by the detensioning operation that took place after 178 days (0.49 years) and involved increasing stress on the strands in order to remove the wedge grips. As such, it is likely that these fractures were eminent irrespective of the additional stress that was applied. Based upon pit depth measurements on the three fractured wires, μ(CR) was determined as 1.89 mmpy (74 mpy) and σ(CR) 0.38 of the mean. From this and considering, first, the difference in the number of wires (24 stressed outer wires for the tendon at issue compared to 21,384 employed in the model) and, second, that the strands were stressed to 60 percent of GUTS instead of 63 percent yielded a $T_f$ of 0.66 years, which is in good agreement with an assumed value of 0.6 years. Further, in all likelihood, the UTS for these wires was greater than the modeling value of 1,862 MPa (270 ksi), as indicated by the fracture stress for the uncorroded wire in Figure 6.2, in which case agreement would probably be even closer. That the remaining wires on this strand did not fracture once three wires broke, as projected to occur according to the analysis assumptions described above, is attributed to the development length for stress transfer to remaining wires being relatively long compared to the length of the tendon, the latter being 3.05 m (10.0 ft).[7,8] Consequently, the unfractured wires likely were stressed in displacement control.

The second model qualification opportunity was based upon pit depth measurements performed on strands from one of the two failed tendons recovered from the Ringling Causeway Bridge mentioned above and which were disclosed at a bridge age of approximately eight years. Figure 6.7 shows a photograph of sections of this tendon that were acquired from the vicinity of the failure and Figure 6.8 of wires from one strand after separating them and cleaning. Based

*Figure 6.7* Photograph of as-received sections of two failed tendons.

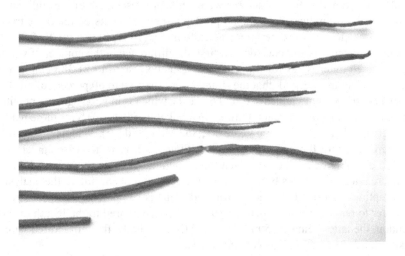

*Figure 6.8* Photograph of wires from a fractured strand subsequent to cleaning.

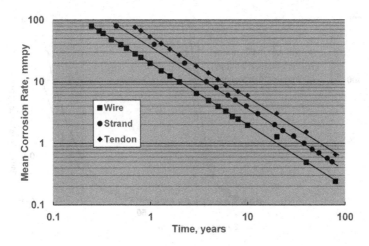

*Figure 6.9* Plot of $T_f$ for wires, strands, and tendons as a function of $\mu$(CR).

upon the measured pit depths and inputting them into the tendon failure model described above, $T_f$ was calculated as 4.2 years for one strand and 3.3 years for the other. Of course, it is not known when during the assumed 7.5 years that these particular fractures occurred; and they may even have resulted from tensile overload as a consequence of contiguous strand fractures. Also, in all likelihood, corrosion continued and pit depths increased further subsequent to the wire fractures. On this basis, the actual times of fracture may have been greater than the calculated value. Irrespective of this gap, the results are generally consistent with time at which the tendon failures were reported.

Based upon the above fracture/failure projection methodology, Figure 6.9 then provides a plot of $T_f$ for wires, strands, and tendons as a function of $\mu$(CR) with a $\sigma$(CR) of 0.5.[9] Considering that wire fractures have been reported as having occurred in less than one year, as noted above for the FHWA mock-up tendon, mean corrosion rates in excess of 20 mmpy (790 mpy) occurred according to the Figure 6.9 data. However, the Figure 6.10 results are based upon a 162-tendon system, each with 22 strands, stressed to 63 percent of GUTS. Model modification to address a four-tendon system stressed to 60 percent of GUTS yielded a $T_f$ for wires of 0.52 years, which is in excellent agreement with what actually transpired, as discussed above. Note also that the Figure 6.9 results project a tendon failure at 75 years, which is the typical design life for major bridge structures, to result from a $\mu$(CR) of less than one mmpy (39 mpy).

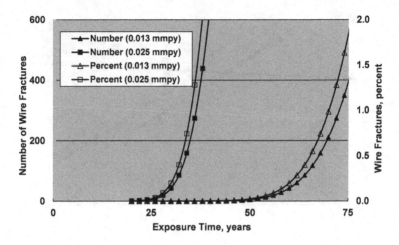

*Figure 6.10* Plot showing $T_f$ and the initial progression of wire fractures per two-year time increment for $\mu(CR)$ = 0.013 and 0.025 mmpy (0.5 and 1.0 mpy) and $\sigma(CR)$ 50 percent of $\mu(CR)$.

## FRACTURE AND FAILURE RATES SUBSEQUENT TO FIRST OCCURRENCE

Of particular importance, in addition to $T_f$, are fracture and failure rates subsequent to an initial occurrence since these define the urgency for repair/replacement. Based upon the above fracture projection methodology, Figure 6.10 shows a plot of model-projected timing for an initial wire fracture and subsequent fracture progression thereafter to 75 years for a system with 162 tendons, each with 22 strands employing two relatively modest $\mu(CR)$ values (0.013 and 0.025 mmpy (0.5 and 1.0 mpy) with $\sigma(CR)$ 50 percent of $\mu(CR)$).[3] Figure 6.11 then illustrates the number and percent of wire fractures per successive two-year time increment. Following from this, Figure 6.12 shows the corresponding results for strand fractures employing the same corrosion rate statistics as for the Figures 6.10 and 6.11 analyses. These results indicate that strand $T_f$ for the lower $\mu(CR)$ (0.013 mmpy or 0.5 mpy) exceeds 75 years but for the higher (0.025 mmpy or 1.0 mpy) is 40 years. Also, tendon failures are shown in Figure 6.13 as nil to 75 years for the lower mean corrosion rate but for the higher 56 years with all having failed prior to 75 years (not shown). Similar to the above representation for the progression of incremental wire fractures (Figure 6.11), corresponding strand fractures per two-year increments are plotted versus time in Figure 6.14 and for tendons in Figure 6.15.

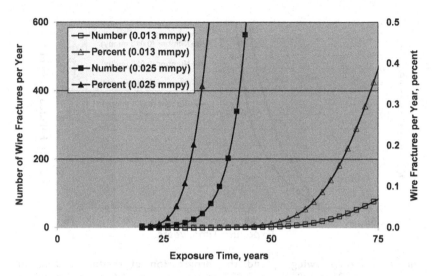

*Figure 6.11* Plot showing T$_f$ and the progression of wire fractures per year for μ(CR) = 0.013 and 0.025 mmpy (0.5 and 1.0 mpy) and σ(CR) 50 percent of μ(CR).

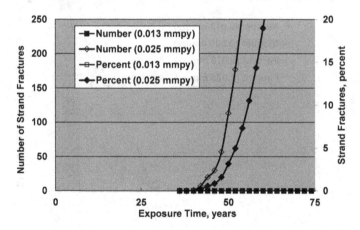

*Figure 6.12* Plot showing T$_f$ and the progression of strand fractures for μ(CR) = 0.013 and 0.025 mmpy (0.5 and 1.0 mpy) and σ(CR) 50 percent of μ(CR).

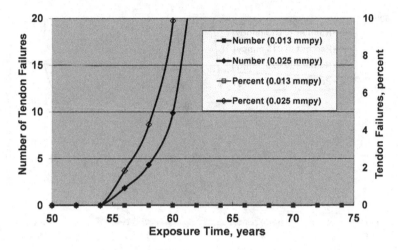

*Figure 6.13* Plot showing T$_f$ and the progression of tendon failures for μ(CR) = 0.013 and 0.025 mmpy (0.5 and 1.0 mpy) and σ(CR) 50 percent of μ(CR).

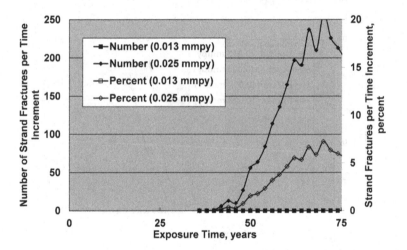

*Figure 6.14* Plot showing T$_f$ and the progression of strand fractures per two-year time increment for μ(CR) = 0.013 and 0.025 mmpy (0.5 and 1.0 mpy) and σ(CR) 50 percent of μ(CR).

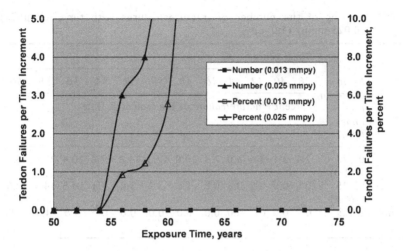

*Figure 6.15* Plot showing T$_f$ and the progression of tendon failures per time increment for $\mu$(CR) = 0.013 and 0.025 mmpy (0.5 and 1.0 mpy) and $\sigma$(CR) 50 percent of $\mu$(CR).

While the number of wire fractures exhibits a smooth increase with time, those for strands and tendons show a scatter about a general trend. This happens because projections for strands and tendons involve grouping, in the first case of successive six random numbers and the latter 22 groups of six, as explained above. Of course, these projections apply only to situations with the indicated corrosion rate statistics. Taking this one step further, Table 6.1 lists percentages of wire and strand fractures and tendon failures from 38 to 60 years for $\mu$(CR) = 0.025 mmpy (1.0 mpy) and shows that tendon failures are initiated at year 56 and have reached almost 10 percent at year 60. However, if these failures are spatially concentrated, for example, on 10 percent of a bridge (similar to what was reported for the Ringling Causeway, as noted above), then essentially all tendons in that local region are projected to have failed by year 60. While this projection might appear to violate the random number generation approach employed in the model, it does not since essentially the same random numbers are involved but are simply relocated in the spreadsheet, thereby representing the localization of material or construction improprieties and a corresponding, locally high distribution of corrosion rate. Note also that it is unlikely that the same numbers will appear in separate runs because in Microsoft Excel there is an equal probability for any number in the selected range appearing in any given cell. Consequently, some numbers may not appear at all and others multiple times. This specific point of localized attack is addressed in more detail subsequently.

*Table 6.1* Listing of the number of fractures/failures with time for μ(CR) 0.025 mmpy (1.0 mpy)

|  |  | Time, years | | | | | | | | | | | |
|---|---|---|---|---|---|---|---|---|---|---|---|---|---|
|  | Mean corrosion rate, mmpy (mpy) | 38 | 40 | 42 | 44 | 46 | 48 | 50 | 52 | 54 | 56 | 58 | 60 |
|  |  | Percent Fracture/Failure at the Indicated Time | | | | | | | | | | | |
| Wires | 0.025 (1.0) | 2.1 | 3.1 | 4.4 | 6.0 | 7.9 | 10.0 | 12.3 | 14.8 | 17.6 | 20.4 | 23.4 | 26.5 |
| Strands | 0.025 (1.0) | 0.0 | 0.0 | 0.2 | 0.6 | 0.8 | 1.6 | 3.2 | 5.0 | 7.3 | 10.5 | 14.3 | 19.0 |
| Tendons | 0.025 (1.0) | 0.0 | 0.0 | 0.0 | 0.0 | 0.0 | 0.0 | 0.0 | 0.0 | 0.0 | 1.9 | 4.3 | 9.9 |

## MODELING VARIABLES

Subsequent to the development of the basic tendon failure model, as described above, focus has been placed upon the effect of different variables, including the following:

1. Analysis variables (different sets of random numbers)
2. Wire/strand/tendon strength and prestress level
3. Number of strands per tendon
4. Tendon length
5. Time dependence of tendon stress
6. Partitioned attack (spatially variable corrosion rate distribution).

Each of these is discussed below.

*Random Number Set*: To investigate this factor, 10 different analyses were performed, each with a different set of random numbers but the same μ(CR) and σ(CR), the objective being to determine any effect of this variable (random number set) on $T_f$. Table 6.2 then lists $T_f$ results as determined for these analyses (different random number sets) for the case of 162 tendons, each with 22 seven-wire strands, where μ(CR) = 0.076 mmpy (3.0 mpy) and σ(CR) 50 percent of the mean. As noted above, only the six outer wires of strands are counted, which yields a total of 21,384 wires or random numbers. Time increment for the $T_f$ wire determinations was 0.1 years and for strands and tendons 0.5 years. The results in Table 6.2 show that, despite the fact that no two random number sets are identical, timing of an initial wire fracture was the same in each of the 10 analyses to the nearest 0.1 years. However, the number of wire fractures in the

Table 6.2 Listing of $T_f$ results for wires, strands, and tendons along with $\mu(T_f)$ and $\sigma(T_f)$ for ten runs each with a different random number set

| Run | $T_f$ (Wire) | $T_f$ (Strand) | $T_f$ (Tendon) |
|---|---|---|---|
| 1 | 20.43 | 35.2 | 51.5 |
| 2 | | 39.0 | 55.1 |
| 3 | | 29.6 | 57.7 |
| 4 | | 31.6 | 53.9 |
| 5 | | 39.1 | 55.9 |
| 6 | | 35.4 | 54.2 |
| 7 | | 38.2 | 55.6 |
| 8 | | 37.3 | 56.5 |
| 9 | | 35.8 | 54.7 |
| 10 | | 30.4 | 53.5 |
| | $\mu$ | 35.2 | 54.9 |
| | $\sigma$ | 3.5 | 1.7 |
| | Median | 35.6 | 54.9 |
| | $\mu+2\sigma$ | 42.2 | 58.3 |
| | $\mu-2\sigma$ | 28.1 | 51.4 |

initial 0.1 year subsequent to $T_f$ (not shown) ranged from 86 to 171, suggesting that if a smaller time increment were employed, a distinction in $T_f$ for wires would result. On the other hand, $\sigma(T_f)$ for strands was approximately 5 percent of the mean and for tendons 3 percent. This variation in $T_f$ between different runs resulted in the former case (strands) because of the successive grouping of six random numbers, which varied from one run to the next, to represent outer wires of individual strands and in the latter (tendons) from 22 groupings of six.

Figure 6.16 then plots strand fractures versus time to ten percent as projected from the above ten analyses. At the extreme, $T_f$ for one run was 12.0 years and for another 14.5 years yielding a range of 2.5 years or almost 20 percent. This suggests that an initial $T_f$ for strands is likely to be shorter the greater the number of runs or, alternatively, the greater the number of strands. Likewise, Figure 6.18 provides a similar plot for tendons and shows that $T_f$ in this case ranges from 18.5 to 20.0 years. Of practical significance is the range for tendon failures shortly beyond $T_f$, since this reflects the urgency for the failure to be addressed. For example, at year 20, the range of projected tendon failures is from 0.6 to over 3 percent. For a 162-tendon bridge, as analyzed above, this translates to one and five tendons, respectively. This range increases further at progressively higher times. The dashed line in Figure 6.16 indicates the mean plus two standard deviations trend for the tendon failure percentage. Given that the data are normally distributed, as discussed earlier, approximately 2.5 percent of tendons in these analyses are projected to fail in conformance with the latter trend. The finding that this closely corresponds to the upper bound for the ten

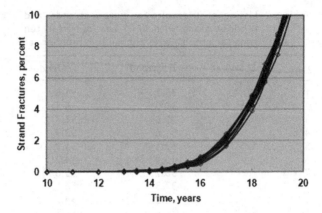

*Figure 6.16* Plot showing onset and subsequent fracture progression for strands as projected by 10 different random number sets with μ(CR) = 0.025 mmpy (1.0 mpy).

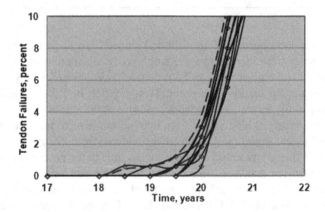

*Figure 6.17* Plot showing onset and subsequent failure progression for tendons as projected by ten different random number sets with μ(CR) = 0.025 mmpy (1.0 mpy).

analyses indicates that the latter trend here should provide a good representation of an extreme or worst-case occurrence. Table 6.3 then lists the percent of failed tendons at $\mu(T_f)$ and $\mu(T_f) - 2 \cdot \sigma(T_f)$ in Figure 6.17. This shows that the extreme of initial failure percentages relative to the mean is greatest at early times.

*Prestress and Wire/Strand/Tendon Strength*: The above analyses all assumed a wire/strand strength (UTS) of 1862 MPa (270 ksi) (see Equation 6.2), which is the minimum specified value[3]; however, the actual strength is likely to be greater

*Table 6.3* Listing of tendon failure percentages at $\mu(T_f)$ and $\mu(T_f)-2\sigma(T_f)$ from Figure 6.18 at the indicated times

| Year | Percent Failures | |
|---|---|---|
| | $\mu(T_f)$ | $\mu(T_f)-2\sigma(T_f)$ |
| 18.5 | 0.06 | 0.42 |
| 19.0 | 0.12 | 0.64 |
| 19.5 | 0.43 | 1.27 |
| 20.0 | 2.16 | 3.83 |

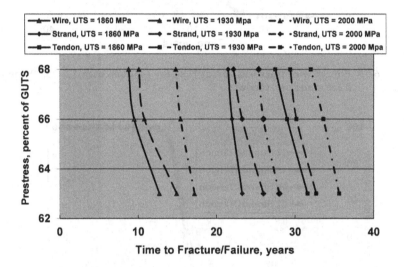

*Figure 6.18* Time-to-failure as a function of prestress for wires, strands, and tendons of three different strengths.

in order to ensure that this criterion is met. As an example of this calculation, note the data point in Figure 6.2 for the test specimen with no pre-corrosion, which failed at 2013 MPa (292 ksi). Figure 6.19 then plots the projected time to an initial failure for wires, strands, and tendons of three different tensile strengths (1862, 1930, and 2000 MPa (270, 280, and 290 ksi)) as a function of prestress. The data indicate approximately a 20 percent $T_f$ decrease for the 1,862 MPa (270 ksi) tendon compared to the 2,000 MPa (290 ksi) one at a prestress 68 percent of GUTS and a 13 percent decrease for prestress 66 compared to 63 percent of GUTS. As mentioned above and explained in greater detail subsequently, prestress decays with time to a long-term, constant value because of progressive concrete creep and shrinkage and strand relaxation.

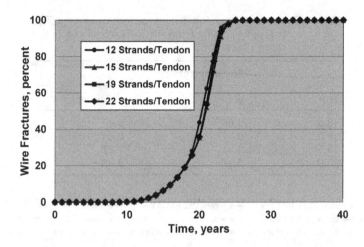

*Figure 6.19* Plot of wire fractures versus time for tendons with 12, 15, 19, and 22 strands per tendon.

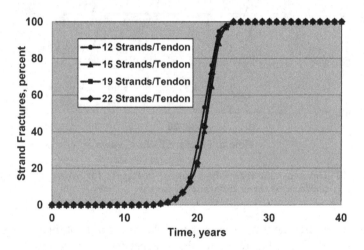

*Figure 6.20* Plot of strand fractures versus time for tendons with 12, 15, 19, and 22 strands per tendon.

*Number of Strands per Tendon*: Analyses in this category considered tendons with 12, 15, 19, and 22 strands with results being shown in Figures 6.19–6.21 as plots of fractures/failures versus time for wires, strands, and tendons, respectively. Differences progressively increase upon proceeding from wires to strands and then to tendons, which is attributed to different random number of sets being employed in each case and grouping of numbers to represent strands and then

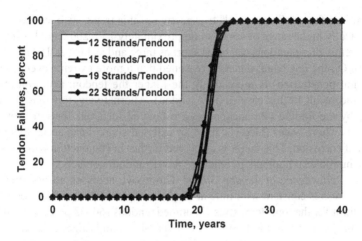

*Figure 6.21* Plot of tendon failures versus time for tendons with 12, 15, 19, and 22 strands per tendon.

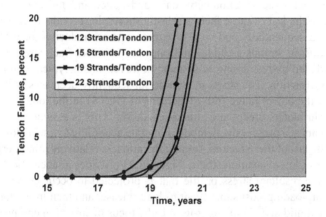

*Figure 6.22* Expanded scale view of the trends from Figure 6.21.

tendons, as discussed above. For wires and strands, these differences at low fracture percentages at and just beyond $T_f$ are relatively minor. However, this is not the case for tendons; and to illustrate this phenomenon, Figure 6.22 shows an expanded scale view of the initial trends from Figure 6.21. Here, failures at year 19 range from none to over 4 percent and at year 20 from almost 4 to over 19 percent. However, these differences are primarily due to different random number sets and not to the number of strands per tendon.

*Tendon Length*: Any effect of this factor (tendon length) upon corrosion and resultant fractures/failures is likely to reflect statistical considerations and the likelihood that a relatively long tendon will exhibit greater localized corrosion section loss than a shorter one.[10] Irrespective of this trend, such tendon lengths are typically in the range of 45–60 m (approximately 150–200 feet); however, in the extreme, these tendons can be as short as three and as long as 150 m (almost 500 ft). On the one hand, it can be reasoned that a more extreme event (deeper corrosion penetration) is more likely the longer the tendon; and so a shorter $T_f$ should result in this case (longer tendon). Alternatively, this may be moderated by the finding that anchorages are favored locations for soft grout and for voids where water (bleed or from an external source) can accumulate with resultant corrosion. This factor is discussed further in conjunction with analyses involving reduced numbers of tendons (fewer random numbers).

*Time Dependence of Tendon Stress*: The above analyses all assumed that tendon stress is constant with time at 63 percent of GUTS,[1,3,9] which reflects an upper limit for the long-term stress in draped tendons (60–63 percent of GUTS) that is projected to result at mid-point of span-by-span bridges, where bending moment in the concrete is greatest. However, strands are typically stressed initially to 80 percent of GUTS. This subsequently decreases upon load release to an assumed value of 70 percent of GUTS as a consequence of seating of wedges in anchorages, friction between strands, duct, and deviators, and elastic shortening. The last of these factors (elastic shortening) results because strands are stressed sequentially and so a progressive stress reduction results in previously stressed strands as additional ones are stressed. Such stress reduction can be reduced by restressing of strands; however, this is typically not done in bridge construction. The stress in strands subsequently decreases with time to a value in the above range (60–63 percent of GUTS) in the long term due to concrete shrinkage, creep, and strand relaxation. This is assumed to transpire over 27 years at a progressively moderating rate according to a semi-logarithmic trend.[11] The last of these factors, strand relaxation, is relatively minor, especially in the case of low relaxation strand. Figure 6.23 provides a schematic representation of the resultant stress profile that is projected to occur along a tendon length upon loading, just subsequent to load release, and then in the long term. The stress at mid-span in the long-term is the focus of structural engineers since bending moment in the concrete is greatest here; however, tensile stress in tendons is slightly higher to the loading end side of this, as shown schematically in Figure 6.23, and is estimated as 65 percent of GUTS. Figure 6.24 illustrates this tendon stress (TS) reduction with time (T) schematically to long-term values of 60, 63, and 65 percent of GUTS. The equation for stress decay to each of the three long-term values cited above is given as:

$$TS = -0.49 \, LOG(T) + 66.62 \quad \text{(long-term stress 65\% of GUTS)} \tag{6.5}$$

$$TS = -0.67 \cdot LOG(T) + 65.26 \quad \text{(long-term stress 63\% of GUTS)} \tag{6.6}$$

$$TS = -0.98 \, LOG(T) + 63.23 \quad \text{(long-term stress 60\% of GUTS)} \tag{6.7}$$

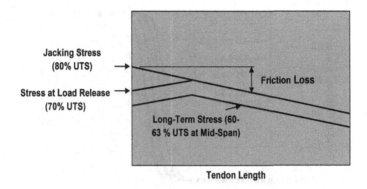

*Figure 6.23* Schematic representation of the stress profile along a tendon both initially and in the long term (stressing end is at the left).

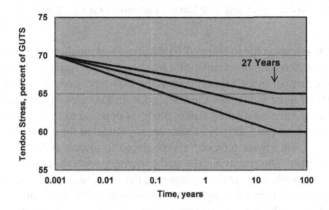

*Figure 6.24* Schematic illustration of tendon stress decrease with time due to concrete creep, shrinkage, and strand relaxation.

To incorporate this time dependence into the above fracture/failure projection protocol, the constant term in Equation 6.2 was replaced by one of the three above equations according to value for the long-term stress being considered (60, 63, or 65 percent of GUTS). By so doing, the fracture/failure analysis protocol was modified to incorporate this stress time dependence. This was done for, first, situations representative of soft grout where tendon failure can occur within several years of construction (Case 1) and, second, where failure occurs several decades post-construction as a consequence of entry of water into grout voids or presence of bleed water (Case 2).[12]

*Case 1*: Figure 6.25 shows analysis results for the percentage of tendon failures versus time where $\mu(CR) = 0.76$ mmpy (30 mpy) and $\sigma(CR)$ is 50 percent of

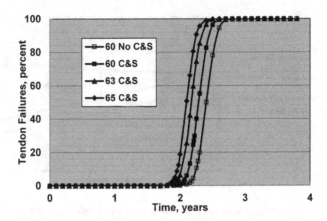

*Figure 6.25* Plot of tendon failure percentage versus time for four different stress situations and Case 1 corrosion rate statistics.

the mean based upon the average of 10 runs (random number sets). This μ(CR) value is considered representative of a harsh soft grout situation, as explained earlier. The number in the caption reflects the long-term stress that is reached at year 27 (see Figure 6.24); and the C&S notation indicates that concrete creep and shrinkage have been taken into account, as opposed to tendon stress being constant with time (no C&S). All trends are sigmoidal as for the above analyses (for example, see Figure 6.6) and are displaced to greater time the lower the long-term stress. The results for the analysis where concrete creep and shrinkage are not considered (stress is constant at the long-term value) are displaced to greater time than for the analyses that incorporate C&S. While the four trends are spaced relatively close to one another time wise, still the distinctions that do exist can be significant. For example, Figure 6.26 provides an expanded scale view of the early failures from Figure 6.25 and shows that at year two the 60 C&S data project an initial tendon failure (0.55 percent failure); however, the 63 C&S results show 6 percent failure at this same time and the 65 C&S trend almost 20 percent failure. The dashed line indicates the μ+2σ on failure percentage for the 65 C&S trend, thereby providing a worst-case projection for this stress state.

*Case 2*: Figure 6.27 then shows similar results for μ(CR) = 0.076 mmpy (3.0 mpy) and σ(CR) 50 percent of the mean, reflecting a situation where bleed water or water from an external source is present within a grout void. The trends here are similar to those for Case 1 (Figure 6.26) but are displaced to greater time; however, the 60 C&S data almost superimpose upon the No C&S results. This occurred because beyond year 20 most of the concrete creep and shrinkage has

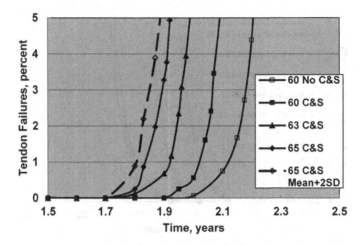

*Figure 6.26* Expanded view of early tendon failures from Figure 5.26.

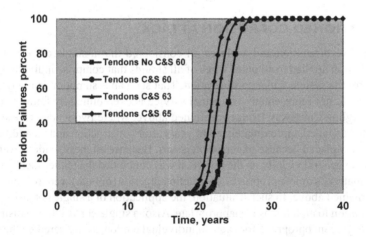

*Figure 6.27* Plot of tendon failure percentage versus time for four different stress situations and Case 2 corrosion rate statistics.

already transpired (see Figure 6.24). Figure 6.27 provides an expanded scale plot of the initial data from Figure 6.27 and shows that at year 20.5 0.3 percent tendon failures are projected for the 60 C&S case, while almost 4 percent failure transpire for the 63 C&S data and 16 percent for the 65 C&S. Also included is the trend line for the mean plus two standard deviations of the 65 C&S results. As noted above, this is projected to reflect the most corroding 2.5 percent for those results and, as such, a worst-case extreme.

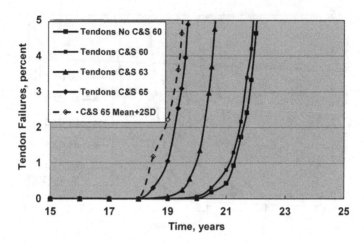

*Figure 6.28* Expanded scale plot of the tendon failure percentage versus time results from Figure 6.27.

## PARTITIONED CORROSION ATTACK

The above discussion and analyses all considered that a single corrosion rate distribution applied to all outer wires of all strands that comprise individual tendons for the system under consideration. That such an assumption is likely simplistic and not representative of actual service is exemplified by findings from the Ringling Causeway Bridge mentioned above, where two tendons were discovered as failed approximately eight years after construction and an additional 15 were replaced because of corrosion issues. Because all these tendons were in the same general vicinity on the bridge, it was concluded that either locally deficient materials or inappropriate construction practice (or both) were responsible, as explained above. In such a situation, the application of a single corrosion rate distribution to all tendons is inappropriate. Also, a single corrosion rate distribution may be inappropriate for even an individual tendon, as indicated by the section view of a failed tendon from the Ringling Causeway Bridge in Figure 3.5, where regions of wet plastic (soft) grout, white chalky grout, sound gray grout, and a void that contained free water were identified. Each of these two situations (corrosion limited to, first, a subset of tendons and, second, a single tendon or a portion thereof) is addressed below.

## SYSTEM WITH SIX CORRODING TENDONS

*General*: Analyses in this category pertain to situations where corrosion is limited to six tendons, irrespective of how many tendons comprise the bridge.

These could be on a single span as noted above. For a tendon configuration as in Figure 2.3, loss of a second tendon would reduce the structural capacity by about one-third; however, if these two failed tendons are on the same side of the span, then excessive deflection and concrete cracking would result; and the bridge would be closed for immediate repairs, as noted above. Examples of both Case 1 and Case 2 situations are analyzed below.

*Case 1 ($\mu$(CR) = 0.76 mmpy (30 mpy) and $\sigma$(CR) 50 percent of $\mu$(CR))*: Figure 6.29 illustrates the results of analysis where only wires on six tendons are corroding. This reveals that fracture progression for wires is smooth up to the time when strand fractures commence, beyond which small irregularities result. Strand fractures, once initiated, also exhibit small irregularities; however, upon the onset of tendon failures, both wire and strand fracture percentages necessarily exhibit sharp increases.

An alternative approach is to average the results from a number of analyses, all with the same distribution of corrosion rate, as shown in Figure 6.30 for 10 runs with the same number of corroding tendons (six) and corrosion rate statistics as in Figure 6.29. Doing this results in much reduced irregularities. Figure 6.31 then plots the timing of an initial tendon failure for 10 individual runs (different sets of random numbers) considering five different possibilities for the percentage of tendons that are corroding. Unless indicated otherwise, there are 12 strands per tendon. That no tendon failures are projected in the case where 25 percent of the 12 strands are corroding resulted because with 12

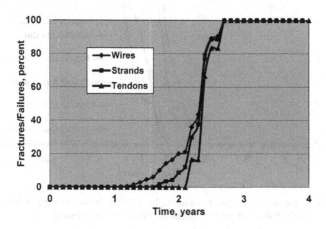

*Figure 6.29* Plot of projected wire and strand fractures and tendon failures with time for a system comprised of six corroding tendons.

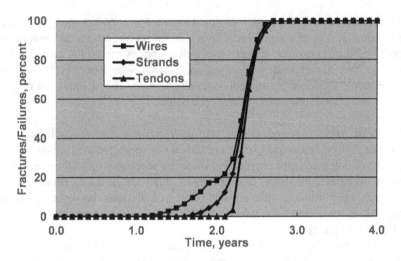

*Figure 6.30* Plot of fracture/failure onset and progression with time for the average of ten runs with the same corrosion rate statistics as in Figure 6.29.

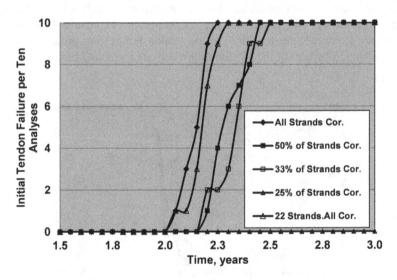

*Figure 6.31* Time of an initial tendon failure determined for each of ten runs with different percentages of strands corroding.

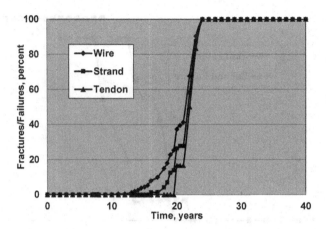

*Figure 6.32* Example fracture/failure results for an analysis involving Case 2 corrosion rate statistics.

strands per tendon, four are required to fracture in order for the GUTS to be reached or exceeded in those remaining. The initial failure trends for the two cases where all strands are corroding (12 strands per tendon in one case and 22 in the other) are essentially the same, apparently because the population of random numbers in the two instances was sufficiently large that the resulting extreme event in each case was essentially the same. However, the time for an initial failure in situations where 50 and 33 percent of 12 strands are corroding, while also essentially the same, are displaced to greater time. This reflects the smaller random number population and, consequently, a reduced likelihood of an extreme event (higher $T_f$).

*Case 2 ($\mu(CR)$ = 0.08 mmpy (3.0 mpy) and $\sigma(CR)$ 50 percent of $\mu(CR)$*: Similar analyses as above were performed based upon Case 2 corrosion rate statistics, with example results for onset and fracture/failure progression with time for wires, strands, and tendons being shown in Figure 6.32. The trends here are generally similar to those for Case 1 (Figure 6.29) but are displaced to greater time. As for Case 1, averaging results from 10 analyses (not shown) resulted in generally smooth fracture/failure trends, similar to the results in Figure 6.30. Figure 6.33 then shows a plot of failures versus time for both an initial and second tendon based upon ten Case 2 analyses and indicates that the second failure lagged the first by from 6 to 18 months. The timing of a second failure can be critical to bridge structural integrity, as explained above. Such occurrences were not addressed for Case 1 situations because the time lag for a second failure at such a high corrosion rate is relatively small.

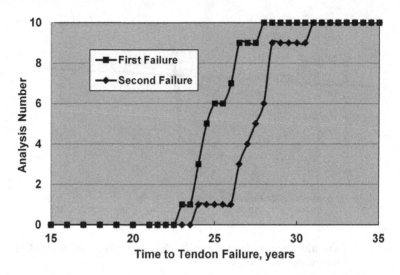

*Figure 6.33* Plot showing time of an initial and second tendon failure based upon ten analyses for Case 2 corrosion rate statistics.

*Table 6.4* Listing of analysis parameters for analyses investigating failure trends for systems with different numbers of tendons

| Analysis | Number of corroding tendons | Strands per tendon |
|---|---|---|
| 1 | 6 | 12 |
| 2 | 12 | 12 |
| 3 | 24 | 12 |
| 4 | 162 | 19 |

## NUMBER OF CORRODING TENDONS

Analyses were performed that project fracture/failure percentages for the four corroding tendon configurations listed in Table 6.4 with results being as shown in Figure 6.34. It is unclear if there is any distinction in results for the three higher number of tendon cases, suggesting that the population of random numbers employed for the 162 tendon analyses did not result in a more extreme event than for those involving just 12 tendons. On the other hand, the $T_f$ trend for the six tendon analyses is displaced to greater time by from one to two years (5–10 percent), indicating the reduced likelihood of an extreme event compared to results with a greater number of tendons.

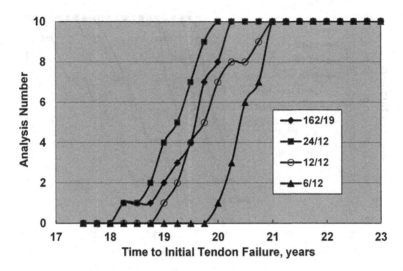

*Figure 6.34* $T_f$ results for ten analyses employing the tendon configurations
listed in Table 6.4.

## TENDONS WITH FIFTY PERCENT OF STRANDS CORRODING

Analyses in this category involved the same tendon configurations as in
Table 6.4 and Figure 6.35 but with only wires on strands comprising 50 percent
of tendons corroding; that is, having an assigned random number. Figure 6.36
provides results for these analyses in comparison to those in Figure 6.34 and
shows that with wires on one-half of tendons corroding $T_f$ is displaced to greater
time and with reduced slope (greater variability), both presumably due to the
reduced population of random numbers and, hence, a lower probability of an
extreme event.

## TENDONS WITH WIRES CORRODING ACCORDING TO TWO DISTRIBUTIONS

As noted above in conjunction with Figure 3.5, more than one corrosion rate distribution can result locally on the same tendon. Also, only a fraction of strands
may have corroding wires, as discussed in conjunction with Figure 6.36. To
address a situation that involves a combination of these two factors, three sets
of 10 analyses were performed on a system comprised of six tendons, each with

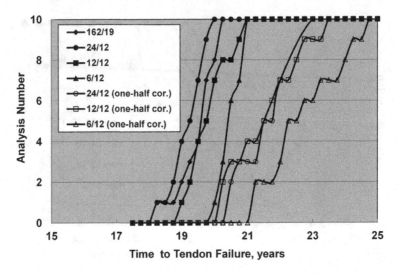

*Figure 6.35* Plot of tendon T$_f$ for ten analyses showing results with 50 percent of strands having corroding wires in comparison to the Figure 6.34 results.

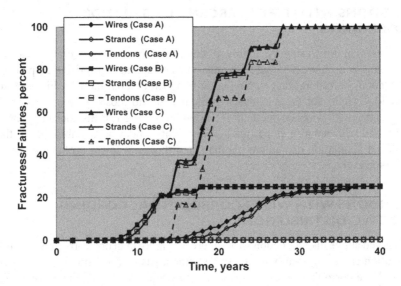

*Figure 6.36* Plot of wire/strand fractures and tendon failures with time as determined by analyses for Cases A, B, and C defined above.

12 strands. In one set, termed Case A, $\mu(CR)$ for wires on three of the 12 strands was 0.076 mmpy (3.0 mpy), and for a second set (Case B) $\mu(CR)$ for wires on three strands was 0.152 mmpy (6.0 mpy) with $\sigma(CR)$ 50 percent of the mean in both instances. In the two analyses, no corrosion was assumed for wires on the remaining nine strands of each of the six tendons. For a third situation (Case C), analyses were performed where wires on three of the 12 strands corrode at the lower of the two above mean rates (0.076 mmpy) and for a second three at the higher (0.152 mmpy). Figure 6.36 shows the results for each of these three situations. Thus, for the three strands corroding analyses, $T_f$ is shorter and wire and strand fracture progression, once commenced, is greater for the higher mean corrosion rate (Case B), with all wires on these three strands having fractured after 18 years. For the lower corrosion rate (Case A), this time for all wires on corroding strands to have fractured is extended to 37 years. No tendons are projected to fail in either Case A or B, since for tendons with 12 strands this requires fracture of a fourth strand; however, only three are treated as corroding. For Case C (tendons with six of the 12 strands corroding, three each at one of the above two mean rates), initial wire and strand fractures superimpose upon the Case B results, since no fractures have yet transpired for Case A wires. However, once Case A wire and strand fractures commence, because six of the 12 strands are corroding in Case C, all tendons are projected to have failed by year 27, along with all wires and strands necessarily having fractured also by this time.

## EFFECT OF MEAN CORROSION RATE

Little information is available regarding in-situ $\sigma(CR)$ values for corroding bridge tendons; however, a value of 0.38 was determined for corroded wires on the failed FHWA mock-up tendon discussed above.[1] Also, values of but 0.08–0.17 and 0.06–0.14 were reported for two strands from the Ringling Causeway Bridge that were evaluated (see Figures 6.8 and 6.9).

The above analyses all employed a $\sigma(CR)$ of 50 percent of $\mu(CR)$. To investigate the effect of this parameter upon failure projections, analyses were also performed employing $\sigma(CR)$ values (designated as "s" in the figure caption) of 0.3, 0.4, and 0.5 of $\mu(CR)$ with results being shown in Figure 6.37. This indicates $T_f$ to be displaced to greater time by about 10 percent for the lowest compared to highest of these $\sigma(CR)$ values. Assuming that a $\sigma(CR)$ of 0.5 represents an upper limit to what might occur in actual structures, the above analyses are projected to capture worst-case occurrences.

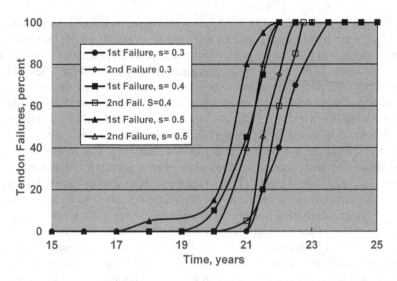

*Figure 6.37* Percentage of tendon failures versus time employing Case 2 μ(CR) and σ(CR) (s in the caption) of 0.3, 0.4, and 0.5 (average of 10 analyses each).

## REFERENCES

1   W.H. Hartt and S-K. Lee, "Corrosion Forecasting and Failure Projection of Post-Tension Tendons in Deficient Cementitious Grout," Report No. FHWA-HRT-17-074, Federal Highway Administration, Washington, DC, April 2018.

2   W.H. Hartt and S-K. Lee, *Corrosion*, Vol. 72, 2016, p. 991.

3   W.H. Hartt and S-K. Lee, *Corrosion*, Vol. 74, 2018, p. 241.

4   *Structure Design Manual*, Vol. 1, Section 4.5.5, Florida Department of Transportation, Tallahassee, FL, 2019.

5   A-L. Stauder, and W.H. Hartt, *Cathodic Protection of Pre-Tensioned Concrete: Part I – Brittle Fracture Propensity of Corrosion Damaged Prestressing Tendon Wire*, paper no. 635 presented at CORROSION/98, NACE (Houston), 1998

6   W.H. Hartt, A. Poeydomenge, A-L. Stauder, and W.T. Scannell, "Long-Term Effects of Cathodic Protection on Prestressed Concrete Bridge Components," Report No. FHWA-RD-98-075, Federal Highway Administration, Washington, DC, 1998.

7   C. MacDougall and F. M. Barlett, *ACI Structural Journal*, Sept.–Oct., 2003, p. 581.

8   C. MacDougall and S. Li, *PCI Journal*, Sept.–Oct, 2007, p. 96.

9   W. H. Hartt, *Corrosion*, Vol. 74(7), 2018, p. 768.

10  W.H. Hartt, *Corrosion*, Vol. 75, 2019, p. 1146.

11  G. Troxell, J. Raphael, and R. Davis, "Long-Term Creep and Shrinkage Tests of Plain and Reinforced Concrete," Proc. ASTM Workshop on the Effects of Aging, Philadelphia, PA, July 6-8, 1994.

12  W.H. Hartt and T.S. Theryo, *Corrosion*, Vol. 76, 2020, p. 1088.

# Chapter 7

# Inspection, condition assessment, and remediation options

The failure projection methodology, discussed in Chapter 6, is fine as a forensic tool; however, its application to forecasting requires that the present residual corrosion state of tendons be quantified. The options for accomplishing this are, first, real-time monitoring and, second, periodic condition assessment. If determined necessary, either of these could be followed by corrective intervention. Options for each of these two categories are listed and discussed below.

The only real-time monitoring option for grouted tendons, other than remote monitoring via on-site video, which would apply to external tendons only and is likely unrealistic, is acoustic emission (AE). Visual inspection also qualifies as a means for condition assessment, as indicated below. Unless invasive measures are taken, however, visual inspection is applicable to external tendons only and is limited to determining whether or not failure has occurred, as evidenced by its sagging or having separated (see Figure 3.7 as an example). In the case of internal tendons, visual inspection requires excavation of concrete to expose a section or sections of duct and subsequently removing a length of duct and then grout, which is likely to only be done locally. Also, the quality of repair subsequent to such an inspection can be problematic. Duct removal and grout excavation can also be performed on external tendons. Acoustic emission instrumentation, on the other hand, senses the sound wave that accompanies any wire and strand fracture(s) and can be employed for both internal and external tendons. It is unlikely, however, that AE would be used unless there are prior concerns or indications of a potential problem. The UK Highways Agency (now Highways England) published Bridge Advice Note BA 86/06, which describes acoustic emission for tendons, as well as a bibliography pertaining thereto.[1] This was subsequently updated as CS 464.[2] In addition, Hamilton et al.[3] have demonstrated that monitoring for fractures and failures can be accomplished in the case of unbonded tendons by strain gauges mounted upon anchorages.

Existing options with regard to condition assessment are the following:

1. Visual inspection (see above)
2. Vibration testing

DOI: 10.1201/9781003328193-7

3. Electrochemical impedance spectroscopy (EIS)
4. Ground penetrating radar (GPR)
5. Impact echo (IE)
6. Borescope inspection
7. Magnetic flux leakage/magnetic main flux testing.
8. Electrically Isolated Tendon (EIT) as described in Chapter 4 for PL3.

Options 2 and 3 are applicable to external tendons only, whereas the option 4 has been employed for both tendon types. Vibration testing has been used on a limited basis and involves striking a tendon of interest and measuring natural frequency of the resulting vibration, recognizing that for a linear system this is a function of mass and tension, the latter being reduced if one or more strands have fractured.[4] The technology has been applied to qualify external tendons that are in the vicinity of a failed one, given that corrosion-caused strand fractures are often localized, as discussed above.

To-date, electrochemical impedance spectroscopy has been limited to proof of concept and determination that it can differentiate between active and passive PT strand corrosion. The technology has yet to transition from laboratory to general field deployment.[5]

Ground penetrating radar (GPR) instrumentation consists of a sending antenna that transmits electromagnetic waves with a frequency typically in the range 10 MHz to 2.6 GHz and a receiving antenna. The technology can detect discontinuities or interfaces between materials of different permittivity, such as cracks and delaminations in grout. It cannot, however, identify corrosion or products thereof.

With impact echo (IE) technology, a stress (sound) wave is generated on a duct by surface-mounted instrumentation and this propagates through the grout but is reflected back by internal separations such as cracks. However, like GPR, it cannot detect corrosion per se or corrosion products.

Borescope inspection utilizes a flexible fiber optics cable that transmits illumination with an eyepiece or display at the forward end and an objective lens at the other such that strands within air voids of otherwise grouted ducts can be visually inspected and condition qualitatively assessed.

Although the technology may still be evolving, magnetic flux leakage testing holds considerable promise as an NDT tool for PT systems. This operates on the principle that when a linear steel system, such as strands of a tendon, is magnetized to saturation, the resultant magnetic flux is proportional to metal cross-section area and consequently is reduced in proportion to the extent of any corrosion loss and wire and strand fractures at a given location. Figure 7.1 provides a schematic illustration of this instrumentation in place on a tendon; and Figure 7.2 schematizes the principle of operation whereby a magnetic flux is induced into the strands, the magnitude of which is proportional to the rate at which the flux changes due to any localized reduction in cross-section as

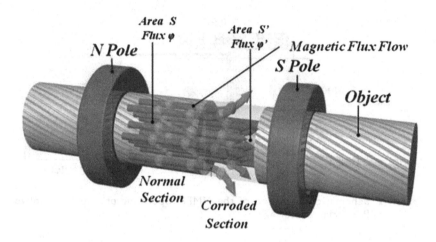

*Figure 7.1* Illustration of the MMFM instrumentation about a tendon and the flux leakage that it detects at a location of wire/strand corrosion loss.

*Source: © Tokyo Rope Manufacturing Co.*

the device is moved along a tendon.[6] Initial instrumentation was based upon what has been termed the solenoid-type main magnetic flux method (MMFM).[7] Figure 7.3 provides a photograph of a MMFM scan in progress on a bridge tendon. This instrumentation has the advantage of requiring relatively little field setup time but is applicable to external tendons only. Also, it has been reported as providing false-positives when cross-section loss is less than about 3 percent.[8] In response to this limitation, what is termed the return flux method (RFM) instrumentation has been developed. Although this requires greater setup time than does the MMFM, it overcomes the small section loss limitation mentioned above. Further, it has been reported that section losses in excess of 15 percent can be reliability detected for internal tendons with 10 or more strands and concrete cover less than 18.8 cm (7.4 in.), even in the presence of vertical conventional reinforcement at 15 cm. (6 in.) minimum spacing. This section loss detection limit reduces to 9 percent in the absence of conventional reinforcement. However, neither MMFM nor RFM is effective for section loss detection within deviators, thick diaphragms, or anchorages. Alternative instrumentation based on this same principle has been developed in Europe and termed the remnant-magnetism-method (R-M-method).[9]

In view of the tendon corrosion issues described and discussed above, emphasis has also been placed in recent years upon design for integrity, inspect ability, and remediation. With regard to design, FDOT now requires for new construction that an additional tendon be included beyond what is structurally necessary. Consequently, bridge shutdown need not occur if a single failed tendon is disclosed. Separately, a deviator redesign has evolved that facilitates

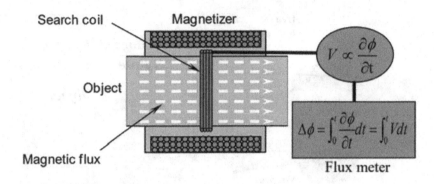

*Figure 7.2* Schematic illustration of the MMFM operating principal.[6] © Tokyo Rope Manufacturing Co.

*Figure 7.3* Photograph of a MMFM scan in progress on a bridge tendon.[7]

tendon removal and replacement should this become necessary, as noted in Chapter 2. With previous designs, such removal can be dangerous to personnel if the tendon or a portion thereof remains under stress. A recent FHWA report addresses various issues associated with PT system components, design issues, and tendon installation and replacement procedures.[10] Also, some states require that anchorages have a port or ports that allow for internal borescope inspection and vacuum grouting of any voids should this be necessary, as also discussed in Chapter 2.[11] Further, what are termed flexible fillers (petroleum wax, grease, or gel) have been employed in lieu of grout for some time on PT bridges in Europe, as well as for nuclear reactor containment vessels in the United States; and bridge projects have recently been completed in the United States using this technology. A comprehensive study by Hamilton et al.[12] addressed various structural and construction issues associated with flexible fillers. An advantage of tendons that utilize flexible fillers, in addition to avoidance of corrosion issues that have been encountered with cementitious grout, is that they are more easily replaced compared to grouted ones should this become necessary. Also, because such tendons are unbonded, it has been shown that wire and strand fractures can be detected by strain gauge instrumentation of anchorages, as noted above. This provides an additional option for real-time monitoring.

A potential negative aspect of tendons with flexible fillers is susceptibility of wires and strands to microbiologically induced corrosion (MIC). In this regard, Little et al.[13–15] reported that failure of a greased mono-strand tendon in a high-rise building was a consequence of corrosion and stress corrosion cracking resulting from moisture ingress that facilitated fungi metabolic processes for which the grease served as a nutrient. The resulting products were acidic and, hence, corrosive, thus leading to the failure. More recently, Presuel-Moreno et al.[16] conducted both outdoor and indoor exposures of steel wire and strand specimens coated with different microcrystalline waxes and sprayed with water with and without different fungi species. In general, the fungi-exposed specimens exhibited greater corrosion than the controls, the extent of which varied for the different fillers.

If the tendon failure projection approach described in Chapter 6 is to be employed for remaining life assessment of actual structures, then a technology is required that quantifies the extent of any existing or present corrosion loss and wire/strand fractures. Of the above assessment alternatives, vibration testing and MMFM/RFM are presently the most likely options for accomplishing this in the case of grouted tendons. In this regard, the model projections reported in Chapter 6 (for example, see Figures 6.6, 6.29, 6.30, and 6.32) suggest conservatively that a minimum of approximately 15 percent wire fractures need to occur prior to the onset of tendon failure. Consequently, if technologies such as vibration testing, MMFM, and RFM are to be employed for tendon qualification, they should be capable of and calibrated accordingly to identify the above threshold (15 percent wire fractures) as being critical. However, as explained subsequently, the percentage of wire fractures to cause tendon failure varies over a relatively wide range depending upon how the fractures are distributed across the wire population that comprises the strands. Consequently, identifying a unique threshold as critical is only appropriate if it is conservative and encompasses all scenarios that can result in failure.

A seemingly feasible but yet apparently underutilized tendon condition assessment methodology involves electrochemical (corrosion) potential measurement. This would be applicable to external tendons only unless there is invasive concrete excavation to expose one or more tendons of interest. The method would be limited to determining if water is present in voids at high elevation along the tendon profile and if the strands being sensed are passively or actively corroding. By this, a 10–15 mm (0.4–0.8 in.) diameter hole is drilled through the duct at a relatively high elevation on draped tendons (shortly inboard of an anchorage, for example) to expose either grout or a grout void. If an air void is exposed, then the procedure would be repeated at lower elevation(s) until grout is encountered. Once this is realized, a connection would be made between a reference electrode mounted upon the grout and the positive terminal of a multimeter with the negative terminal connected to the anchorage. In the case of electrically isolated tendons, the negative terminal would necessarily have to be connected sequentially to each individual strand. Depending upon value of the measured potential, strands could be classified as either actively or passively corroding.

## CORROSION INTERVENTION AND REMEDIATION

Cathodic protection (CP) has been employed for corrosion control of conventional and prestressed reinforcement in concrete. An exploratory feasibility study indicated applicability of CP to PT tendons subject to certain limitations;[17] and a follow-on analysis presented possible design concepts and identified various issues that need to be addressed.[18] Also, a companion technology termed cathodic prevention, which involves reduced cathodic polarization and, hence, less chance of hydrogen embrittlement, was also investigated. Also, high-strength duplex stainless steels have been considered as candidates for strand in prestressed concrete; but there has apparently been no consideration of these for post-tensioning applications, as noted previously.[19,20]

A remediation technology that has evolved over the past several years is tendon inhibitor impregnation. This has typically been performed by removing the cap on tendon anchorages and forcing a low-viscosity corrosion inhibitor under pressure from one anchorage along strand interstices until it emerges at the opposite anchorage.[21–23] Of particular significance is that the inhibitor has been shown to extrude through the outer wire contacts of individual strands and into the surrounding grout, thereby providing protection to the embedded strand outer as well as inner surfaces. Consequently, any actively corroding sites can be repassivated. This has successfully been accomplished for tendons as long as 78 m (256 feet); however, injection can also be performed from one or more intermediate locations, albeit invasively, by accessing the tendon through the concrete cover, should this be deemed necessary.

## THE CHALLENGE OF CONDITION ASSESSMENT

*General*: As indicated above, any approach to tendon condition assessment must necessarily determine the section loss that has transpired at present on corroded and fractured wires and strands. To address this, a recent study[24] provided a methodology whereby this loss, as it exists one and five years prior to tendon failure, designated as $T_f$-1 and $T_f$-5, respectively, can be projected. The first of these two times, $T_f$-1, reflects the likelihood that tendon failure is imminent and, accordingly, that possible lane closure(s) and certainly immediate repair are required and the latter, $T_f$-5, that remedial planning should take place. In all cases, the analyses considered that tendons are comprised of 19 strands each. The subsequent discussion provides findings that determined the percent of wire fractures at these two times for a range of different possibilities.

*System with 162 Tendons*: As noted above, initial analyses were based upon a 162-tendon system, which reflects the design and construction of the Ringling Causeway Bridge. Figure 6.7 provided a plot for such a system of wire and strand fractures and tendon failures with time based upon μ(CR) = 0.08 mmpy (3.0 mpy) and σ(CR) 50 percent of μ(CR). These results indicate that the time of an initial tendon failure is 19.8 years; and the corresponding percentages of wire fractures at $T_f$-1 and $T_f$-5 are 19.4 and 4.2, respectively.

Analyses here consisted of 10 runs, each with a different set of random numbers but the same μ(CR) as above (0.08 mmpy (3.0 mpy)) and σ(CR) values of 0.3, 0.4, and 0.5. Figure 7.4 shows results from these as a plot of the percentage of wire fractures at $T_f$-1 and $T_f$-5 versus $T_f$ for the ten different analyses and reveals that the former (wire fracture percentage) increased in direct linear proportion to the latter ($T_f$) in each case, as explained subsequently. Also, the trends are displaced to lower $T_f$ the higher σ(CR), reflecting the fact that an extreme event is more likely in this case. Regression coefficient ($R^2$) for the different analyses is in the range 0.98–0.99.

*System Comprised of Six Corroding Tendons*: Figure 6.32 presented a plot of tendon failure percentage versus time for a system where but six tendons are corroding according to μ(CR) = 0.076 mmpy (3.0 mpy) and σ(CR) = 0.5. Figure 7.6 then shows the corresponding wire fracture percentages at $T_f$-1 and $T_f$-5 versus $T_f$ for ten runs. Obviously, there is greater scatter in this case compared to the one with 162 tendons (Figure 7.4), with $R^2$ determined for the six corroding tendons case as 0.64, 0.86, and 0.85 for the $T_f$-1 data and 0.65, 0.88, and 0.94 for the $T_f$-5, both for σ(CR) 0.3, 0.4, and 0.5, respectively. At the extreme, the $T_f$–1, σ(CR) = 0.3 and $T_f$-5, σ(CR) = 0.5 data overlap. The greater scatter for the six compared to 162 tendons corroding resulted because of the lesser random numbers for the former and resultant different timing of the irregularities in the fracture/failure trends (Figure 6.31) from one run to the next.

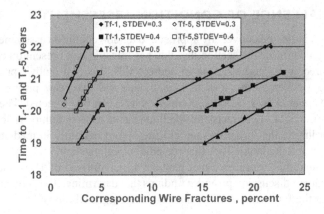

*Figure 7.4* Plot of the percentage of wire fractures corresponding to $T_f$–1 and $T_f$–5 for a 162-tendon system with μ(CR) = 0.075 mmpy (3.0 mpy) and σ(CR) values of 0.3, 0.4, and 0.5.

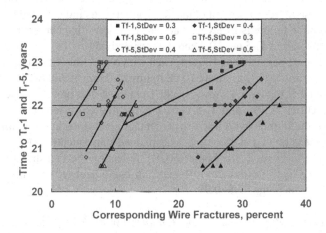

*Figure 7.5* Plot of the percentage of wire fractures corresponding to $T_f$-1 and $T_f$-5 for a six tendons system with all wires corroding according to μ(CR) = 0.075 mmpy (3.0 mpy) and σ(CR) 0.3, 0.4, and 0.5.

*System Comprised of a Single Corroding Tendon*: Examples where but a single failed tendon was disclosed include the Niles Channel Bridge in Florida and the Wando Bridge in South Carolina (Figures 3.1 and 3.7, respectively). To address such a situation, analyses were also performed where all wires of but a single tendon are corroding according to a common distribution of rates. Figure 7.6 shows fracture/failure trends for such a situation according to

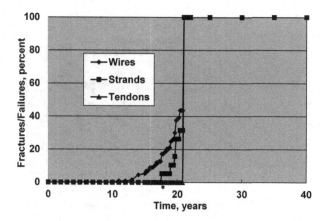

*Figure 7.6* Analysis results for the specific case of a single 19-strand tendon with all wires corroding according to mean rate and standard deviation of 0.075 mmpy (3.0 mpy) and 0.5, respectively.

$\mu(CR) = 0.075$ mmpy (3.0 mpy) and $\sigma(CR)$ 0.5. In this case, $T_f$ is 21.0 years, the greater failure time compared to the 162- and six-tendon examples (Figures 6.6 and 6.29, respectively) resulting from the reduced random number population and corresponding less likelihood of an extreme event. Again, the percentage of wire fractures at $T_f$ is necessarily 100 percent; however, at $T_f$-1 and $T_f$-5, these percentages are 38.3 and 9.0, respectively. In this case, the irregularities or discontinuities in the wire and strand fracture progression prior to tendon failure are more pronounced than for the six corroding tendons example because of the reduced wire population and the relatively large percent increase that accompanies an individual fracture. Figure 7.7 then shows the wire fracture percentages at $T_f$-1 and $T_f$-5 versus $T_f$ for the single tendon corroding case. As for the six compared to 162 corroding tendon examples, the greater percentage of wire fractures at $T_f$-1 and $T_f$-5 in Figure 7.7 reflects the enhanced likelihood that fractures are more evenly distributed across the 19 strands; and, hence, a higher percentage of such fractures at a given $T_f$ is projected at each of these two times ($T_f$-1 and $T_f$-5).

## SINGLE TENDON WITH A FRACTION OF WIRES CORRODING

As noted above, situations may occur where but a fraction of wires on a single tendon is corroding or is corroding according to a greater $\mu(CR)$ than for others. In this regard, Figure 3.5 shows a cross-section of a failed tendon from the Ringling Causeway Bridge; and as discussed in conjunction with this, grout structures include wet plastic grout with void space, segregated chalky grout,

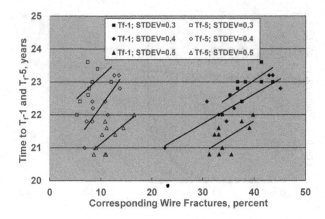

*Figure 7.7* Plot of the percentage of wire fractures corresponding to $T_f$-1 and
$T_f$-5 for a single-tendon system with all wires corroding according to
$\mu(CR) = 0.075$ mmpy (3.0 mpy) and $\sigma(CR)$ 0.3, 0.4, and 0.5.

and sound gray grout. Upon opening the tendon, free water flowed from the void
region. In such a situation, different corrosion rate distributions undoubtedly
apply in each zone. Consequently, analyses were also performed for situations
where only 15, 12, or 9 strands of a single 19 strand tendon exhibit wires that
are actively corroding or are corroding at a greater corrosion rate distribution
than for others. Figures 7.8, 7.9, and 7.10 show results for these three situations,
respectively, in which case $R^2$ values range from 0.09 to 0.85 for the $T_f$-5 results
and 0.08 to 0.67 for the $T_f$-1 but with no discernable correlation with $\sigma(CR)$.
Again, for two of the three analyses (Figures 7.8 and 7.10), the percentage of
wire fractures at $T_f$-1 and $T_f$-5 overlap, thereby reiterating the variability of projections when but a relatively small population of wires is corroding.

Ideally then, any non-destructive testing technology that is to project the present corrosion state of a tendon and forecast the timing of its failure should be
capable of identifying the percentage of local section loss that has transpired at
that time, including that for both fractured and corroded but unfractured wires,
in the analyses considered here at $T_f$-1 and $T_f$-5. However, this percentage of
wire fractures is variable depending upon how section loss and fractures are
distributed across the strands that comprise the tendon, as explained above. The
analysis protocol developed in references 65–67 accommodates this variation
by being based upon a distribution of random numbers, each of which represents an outer wire of a specific strand and with successive groupings of numbers representing a strand that is then assigned to a specific tendon. Thus, the
percentage of wire fractures to cause tendon failure is at a minimum if these

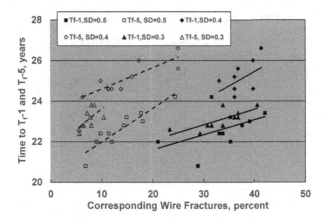

*Figure 7.8* Plot of the percentage of wire fractures at $T_f$-1 and $T_f$-5 in a case where wires of but 15 of the 19 strands of a single tendon are corroding ($\mu$(CR) = 0.075 mmpy (3.0 mpy) and $\sigma$(CR) 0.3, 0.4, and 0.5).

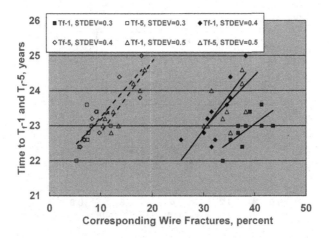

*Figure 7.9* Plot of the percentage of wire fractures at $T_f$-1 and $T_f$-5 in a case where but 12 of the 19 wires on a single tendon are corroding with $\mu$(CR) = 0.075 mmpy (3.0 mpy) and $\sigma$(CR) 0.3, 0.4, and 0.5.

fractures are concentrated on a limited but critical number of strands and larger if these fractures are more distributed. As noted earlier, however, the projection that seven-strand fractures result in tendon failure pertains specifically in this case to 19-strand tendons. The number of strand fractures to cause failure can vary depending upon the number of strands that comprise the tendon.

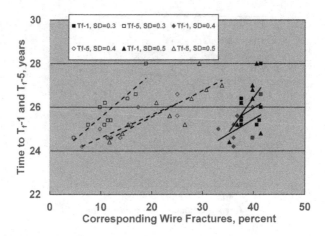

*Figure 7.10* Plot of the percentage of wire fractures at $T_f$-1 and $T_f$-5 in a case where but 9 of the 19 wires on a single tendon are corroding with $\mu(CR) = 0.075$ mmpy (3.0 mpy) and $\sigma(CR)$ 0.3, 0.4, and 0.5.

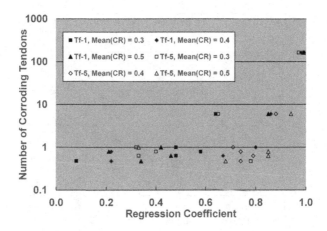

*Figure 7.11* Plot of the regression coefficient versus the number of tendons with corroding wires as determined from the above analyses.

The above results indicate a relatively large variation in the percentage of wire fractures to cause a tendon failure depending upon how such fractures are distributed. As such, this defines an inherent difficulty in relating results from a non-destructive, non-invasive inspection methodology to accurate projection of residual integrity. In this regard, Figure 7.11 provides a plot of the regression coefficient for the above analyses according to the number of corroding tendons.

That some of the tendons are represented in this figure as a fraction reflects that only a portion of the strands thereof have corroding wires. As such, for a single tendon with some or all wires corroding, $R^2$ varies from less than 0.1 to in excess of 0.8. For a non-destructive technology to overcome or address this variability in fracture/failure projection requires that it not only be capable of determining the net amount of any section loss but also how such loss is distributed across the strand population; that is, the extent to which loss is localized as opposed to being distributed. In the absence of such information, a worst-case situation based upon the minimum number of wire fractures to cause most failure can be assumed.

# REFERENCES

1  Highway Structures: Inspection and Maintenance. Inspection. Advice Notes on the Non-Destructive Testing of Highway Structures, DMRB Volume 3 Section 1 Part 7 (BA 86/06), Nine Elms Lane, LondonSW8 5DR, UK.
2  "Non-Destructive Testing of Highway Structures", CS 464, UK Highways Agency (now Highways England), Nine Elms Lane London, England, SW8 5DRUK, September 2019.
3  H.R. Hamilton, J.A. Rice, A.B.M. Abdullah, R. Bhatia, N. Brenkus, and D. Skelton, "Replaceable Unbonded Tendons for Post-Tensioned Bridges," Final Report, Florida Department of Transportation Project No. BDV31-977-15, December 2017.
4  A.A. Sagüés, S.C. Kranc, and T.G. Eason, J. of Bridge Engineering, ASCE, September/October 2006.
5  M. Orazem, D. Blomquist, C. Alexander, "Impedance Based Detection of Corrosion in Post-Tensioned Cables: Phase 2 From Concept to Application," Report No. BDV31-997-35, Florida Department of Transportation, Gainesville, FL., July 2017.
6  S-K. Lee, T. Moriya, H. Itoi, M. Sugawara, H. Kanemaru, "Magnetic Flux-Based Nondestructive Evaluation Technologies for Assessing Corrosion Damage in External and Internal Post-Tensioned Tendons: Development Efforts and Evaluation Result," Draft Report submitted to the Federal Highway Administration, Washington, DC, March 2021.
7  S-K. Lee, T. Moriya, H. Itoi, M. Sagawara, and H. Kanemaru, "Development of Magnetic Flux Based Nondestructive Evaluation Technologies for Detecting Corrosion Damage in External and Internal Post-Tension Tendons," FHWA Draft Report, Federal Highway Administration, Washington, DC, March 2019.
8  S.E. Rehmat, K. Lau, and A. Azizinamini, "Development of Quality Assurance and Quality Control System for Post-Tensioned Bridges in Florida: Case of Ringling Bridge – Phase II," Florida Department of Transportation, Report No. BDV29-997-34, December 2018.
9  B. Hillemeier and A Walther, "Fast Non-Destructive Localization of Prestressing Steel Fractures in Post-Tensioned Concrete Bridges," *Advances in Construction Materials*, Ed. C.U. Grosse, Springler Berlin Heidelberg, New York, 2007, p. 563.
10 T. Ledesma, "Replaceable Grouted External Post-Tensioned Tendons", ´ FHWA-HIF-19-067, Federal Highway Administration, Washington, DC, October 2019.
11 I. Ahmad, N. Suksawang, K. Sobhan, J. Corven, E.A. Sayyafi, S. Pant, and F. Martinez, "Develop Epoxy Grout Pourback Guidance and Test Methods to Eliminate Thermal/

Shrinkage Cracking at Post-Tensioning Anchorages," Report No. BDV29-997-13, Florida Department of Transportation, Tallahassee, FL, January 2016.

12  H.R. Hamilton, J.A. Rice, A.B.M. Abdullah, R. Bhatia, N. Brenkus, and D. Skelton, "Replaceable Unbonded Tendons for Post-Tensioned Bridges," Final Report, Florida Department of Transportation Project No. BDV31-977-15, December 2017.

13  B. Little, R. Staehle, and R. Davis, *International Biodeterioration and Biodegradation,* Vol. 47, 2001, p. 71.

14  B. Little and R. Staehle, *The Electrochemical Society Interface,* Winter 2001, p. 44.

15  R. Staehle and B. Little, "Corrosion ad Stress Corrosion Cracking of Post Tension Cables Associated with Fungal Action," Proceedings Corrosion Research Topical Symposium Microbiologically Influenced Corrosion, NACE International, Houston, TX, 2002.

16  F.J. Presuel-Moreno, C. Castaneda, I. Santillan, A. Kazemi, and F. Tang, "Corrosion Prevention of Bridge Tendons Using Flexible Materials," Final Report No. BDV27-977-10, Florida Department of Transportation, Tallahassee, FL, 2018.

17  J. Bumgardner and A. Sagüés, "Feasibility of Cathodic Protection in Grouted Post Tensioned Tendons – Exploratory Model Calculations," paper no. 7810 presented at COOROSION/16, NACE International, Houston, 2016.

18  A.A. Sagues, I. Lasa, and J. Bungardner, *Materials Performance,* Vol. 59, 2020, p. 24.

19  J. Fernandez, A. Sagues, and G. Mullins, "Investigation of Stress Corrosion Cracking Susceptibility of High Strength Stainless Steels for Use as Strand Materials in Prestressed Concrete Construction in Marine Environments," paper no. C2013-0002686 presented at COOROSION/13, NACE International, Houston, 2013.

20  M. Head, E. Ashby-Bey, K. Edmonds, S. Efe, S. Grose, and I. Mason, "Stainless Steel Prestressing Strand and Bars for Use in Prestressed Concrete, Rept. No. MD-13-SP309B4G, Maryland State Highway Administration, August 2015.

21  D. Whitmore, G. Fallis, H. Liao, S. Strombeck, and I. Lasa, *PTI Journal,* August 2014, p. 17.

22  D. Whitmore, I. Lasa, and L. Haixue, "Impregnation Technique Provides Corrosion Protection to Grouted Post-Tensioned Tendons," *Multi-Span Large Bridges,* paper no. 05008, CRC Press, Boca Raton, FL, 2015.

23  D. Whitmore and I. Lasa, "Investigation and Mitigation of Corroded Bonded Post-Tensioned Tendons," paper no. 10557 presented at COOROSION/18, NACE International, Houston, 2018.

24  W.H. Hartt, "Failure Forecasting of Corroding Bridge Post-Tensioned Tendons," *Corrosion,* Vol. 78(5), 2022, p. 457.

# Chapter 8

# The path forward

A sound, efficient, and reliable infrastructure for the transport of people and materials in a safe, timely, and efficient manner is a requisite for a robust economy and citizen satisfaction. A key contributor toward accomplishing this goal has been the evolution of design, materials, and construction practice for bridges. This is exemplified by post-tensioning, which is at the forefront of bridge materials and design technology and promotes structural efficiency, uses less materials of construction, and provides an esthetically pleasing structure. The concept of segmental PT bridge construction, which facilitates erection over waterways, irregular terrain, and high-traffic volume roadways without disruption, has been a major contributor toward accomplishing this goal. However, as with many civil technological advances, design and construction methods and materials for PT bridges have evolved with time in response to a need to address initially unforeseen issues as these have been identified. Bridge PT tendon corrosion and resultant failures, as these have occurred and discussed in the preceding chapters, have prompted a range of studies and developments regarding materials, design, construction methods, and intervention techniques. Examples include (1) the transition to prepackaged thixotropic grouts, although these are not necessarily without problems, as noted in Chapter 3, (2) flexible fillers, (3) improved grouting practices, (4) enhanced anchorage and deviator designs, and (5) advanced inspection and remediation technologies and techniques. Further advancements will in all likelihood continue to evolve. In this regard, the present state of PT materials, design, inspection, condition assessment, and remediation methods and technologies for bridges likely represents a snapshot in time. A major challenge faced by bridge owners and transportation officials going forward will be managing the vast inventory of structures that predate the above enhancements as these bridges continue to age. As such, corrosion-related issues and resultant tendon failures can only be projected to increase with time. Also, there is no guarantee that "improved" materials, designs, and construction practices will not introduce new, unforeseen problems, as exemplified by the case of thixotropic grout and discussed in Chapter 3. As noted

DOI: 10.1201/9781003328193-8

earlier, the 2006 National Bridge Inventory (NBI) rating for 273 segmental PT bridges was 7.1 (good); however, only 12 years later, tendon corrosion issues had been reported on eight major such structures. Ironically, the grout, duct, and, in the case of internal tendons, concrete cover that are intended to provide corrosion protection also render inspection and condition assessment difficult. There have been additional research projects beyond those listed here that have addressed durability issues for PT bridges, and more will undoubtedly be performed in the future. Collectively, the results from these projects are valuable resources for addressing durability problems. Consequently, a major challenge going forward will be development of improved materials, instrumentation, and techniques and approaches for construction, inspection, preservation, and condition assessment/determination. In the grand scheme, an overall design approach with emphasis on durability starting from conception to final design along with sound detailing practices are required. As such, design engineers must utilize every tool available to best ensure durability and sustainability.

# Index

accelerated corrosión tests 6
acoustic emission (AE) 77
aquaduct 1
arch bridge 1, 2
Arkadiko Bridge 1
ASTM A416 3
ASTM A615 3

Bickton Meadows Footbridge 18
bleed water 38, 45, 65
blister 12
bonded tendons 4
borescope inspection 8, 81

cable stayed bridge 2, 4
cantilever construction 4
cantilever tendon 9
cast-in-place (CIP) PT box girder bridge
    on falseworks 11, 31
cast-in-place segmental balanced
    cantilever bridge 12
clapper bridge 2
concrete creep and shrinkage 7, 46, 61,
    64, 66
Concrete Society 30
continuity longitudinal draped external
    tendons 10
continuity top and bottom tendons 9
corrosion morphology 26
corrosion protection 24, 33
corrosion rate 71–3, 86

deviators 9, 79
diabolos form 9
duplex stainless steels 4

elastic shortening 64
Electrically Isolated Tendon (EIT) 34

Electrochemical Impedence Spectroscopy
    (EIS) 78
electrochemical process 25
environmental cracking 6
Eugene Freyssinet 3
external tendons 9, 77–8
extradosed bridge 2

Federation international du beton (fib)
    30
Finback bridge 2
flexible filler 8, 31, 35, 81, 91
Florida Department of Transportation
    (FDOT) 8, 18, 79
fractured wires 49
fracture projection 54, 65
Freyssinet cone friction PT Anchorage 3
full span bridge construction 14
fully encapsulated 34

gel 4
grease 4
ground penetrating radar (GPR) 78
grouted external tendons 8
grouted internal tendons 8
grouting 6
Guaranteed Ultimate Strength (GUTS) 3
grout segregation 32

high-density polyethylene (HDPE) 8
high strength stainless steels 6
hydrogen embrittlement, 19–20

impact echo (IE) 78
incremental launching bridge 15
internal tendons 77
International Association for Bridge and
    Structural Engineering (IABSE) 31

Laboratoire Central des Ponts des
    Chausees (LCPC) 18, 30
launching nose 15

Magnetic Main Flux Method
    (MMFM)79–81
match-cast joints 13, 31
microcrystalline wax 4
Mid Bay Bridge 18, 27
modified inclined tube test (MITT) 37–9
mono-strand system (MSS) 14–15
Movable Scaffold System (MSS) 14–15

National Bridge Inventory (NBI) 92
New Direction for Florida's PT Bridges
    31
Niles Channel Bridge 18, 37, 84
nuclear power containment 4

overhead gantry 13

pit depth 45–6, 51, 53
pitting 25–6
Plymouth Avenue Bridge 19
polyprophylene (PP) 8
Pont-du-Gard 1
Postbridge in Devon 2
post-tensioning 7–8
Post-Tensioning Institute (PTI) 24–5
precast I girder bridges 15
precast segmental balanced cantilever
    bridge 13, 19, 31
precast segmental span-by-span bridge
    13
prepackaged grout 38
prepackaged thixotropic grouts 91
PT durability issues 31–2

redundant tendon system 33
remnant-magnetism method (R-M
    method) 79
residual fracture stress 47
return flux method (RFM)79, 81
Ringling Causeway Bridge 20–1, 37, 48,
    51, 57, 68, 76, 83, 85

River Schelde Bridge 18, 30
Roosevelt Bridge 19

sail bridge 2
Saint Angelo Bridge 1
segmental bridges 10
segmental concrete bridges 1, 91
segment lifter 13
Selmon West Extension Bridge 19
Service d' Etudes Technique des Routes et
    Autoroutes (SETRA) 18, 30
seven wire spirally strand 4
soft grout 20–1, 37–40, 45, 47, 65–6, 68
span by span 4–5, 23
stand fractures 45, 62, 75
strand relaxation 7, 46, 61
structural protection layers (PL) 34
suspension bridge 2, 4–5

thixotropic grouts 20, 91
time dependence of tendon stress 58, 64
Transverse internal tendons 10
truss bridge 2, 5

UK Department of Transport 31
UK Highway Agency 77
UK Transportation Research Agency 18, 30
unbonded tendons 4, 77
underslung gantry 13
US Federal Highway Administration
    (FHWA) 21–2, 40, 45, 50, 53, 80

vacuum grouting 81
Varina Enon Bridge 19, 23–4, 39
vertical internal tendons 10
vibration testing 77, 81
Virginia Department of Transportation
    (VDOT) 39

Wando River Bridge 19, 84
wire fractures 62

Ynys-Y-Gwas Bridge 30, 32

Zhaozhou Bridge 2

Printed in the United States
by Baker & Taylor Publisher Services

Printed in the United States
by Baker & Taylor Publisher Services